THIS DIRTY WORLD

THIS DIRTY WORLD

Ronald C. Denney, Ph.D.

Nelson

THOMAS NELSON AND SONS LTD
36 Park Street London W1Y 4DE
PO Box 18123 Nairobi Kenya

Thomas Nelson (Australia) Ltd
597 Little Collins Street Melbourne 3000

Thomas Nelson and Sons (Canada) Ltd
81 Curlew Drive Don Mills Ontario

Thomas Nelson (Nigeria) Ltd
PO Box 336 Apapa Lagos

Thomas Nelson and Sons (South Africa) (Proprietary) Ltd
51 Commissioner Street Johannesburg

First published 1971
0 17 138015 0

Printed in Baskerville Linotype
by Hazell Watson & Viney Ltd
Aylesbury, Bucks

To the workers in the fields of
pollution, environment and conservation
whose efforts and achievements are so
frequently ignored.

Forget six counties overhung with smoke,
Forget the snorting steam and piston stroke,
Forget the spreading of the hideous town;
Think rather of the pack-horse on the down,
And dream of London, small and white and clean,
The clear Thames bordered by its gardens green.

William Morris

Plates

Contents

Foreword

by Sir Gerald Nabarro, M.Inst.F., M.P.

Any consideration of man's future must include mastery of pollution. Indeed, this is the epitome of the humanities in the next few centuries. And if man fails to overwhelm the devastating onrush of filth in its diverse material forms, then civilisation ends, and so perhaps life itself. Moreover, population projections are directly related to the extent and degree of pollution.

No more unfortunate decision has been announced by any British Cabinet Minister than Mr Anthony Crosland, M.P., in May 1970 when he said, speaking as Secretary of State for Local Government with responsibilities for the environment, that he was not convinced that there is need for an active Government population policy.

Yet it is almost certain that today's 54 million human beings in Britain will grow to 70 million by A.D. 2000 (thirty years hence and three decades, or just as far forward as the Battle of Britain was ago; a battle millions of us remember as yesteryear) and to more than 100 million in these islands within a century or as far forward as Sedan was ago; and what is one century in the history of man? The same population projectionists say the world's population will *double* by A.D. 2000, and that one in three will be Chinese.

Ponder these statistics. Also, that pollution grows *at least* commensurate to population enlargement.

The question is whether the human race can survive, doubled to 7,000 million people by the dawn of the twenty-first century? Already the waste created as effluent and poison threatens the quality of life in many congested communities; the outflow of nuclear waste and radioactive materials to atmosphere and sea, the disposal of industrial waste, the fouling by emission or accident of oil discharges to beaches and coasts and oil slicks, new phenomena created by such episodes as the *Torrey Canyon* and the *Pacific Glory*: and millions of chimneys, industrial and domestic, pouring poisons into the atmosphere: fourteen million vehicles on our roads today, and twenty million within a decade or two, adding remorselessly to dangers of killer-smogs, disease and human degradation.

At the time I wrote my first Clean Air Bill in 1955, the London smog of three years earlier, when many thousands of people were choked to death, and cattle at Smithfield were asphyxiated, was relatively a recent event. I sought then by Statutory powers a 'revolution at the fireside', to prevent ever again the occurrence of a well-known meteorological phenomenon over a metropolis where toxic products of combustion were belched unremittingly into the atmosphere. The Clean Air Act of 1956, carried through the British Parliament by all-party support, has largely diminished the pollution of the atmosphere by bituminous coal. How slow was remedial legislation in this context will be seen by the fact that it had taken 699 years to legislate: measured from the date that Queen Eleanor was driven from Nottingham by the noxious fumes caused by burning 'Sea-Coales' in the city. Detractors of democracy may point to this effluxion of nearly seven centuries as normal parliamentary progress.

Sadly, fifteen years later I now record that the Clean Air legislation is ailing and failing, because:

1 Supply of solid smokeless fuels does not match demands: notably the much-vaunted 'Brownowski Bullets' manufactured by the National Coal Board. Electricity, gas and oil as alternative fuels do not allow the open fire, so beloved by Britons.

2 Industrial effluents and pollutants, especially alkali and chemical fumes and gaseous outpourings, including notably 'unwashed' power-station smoke, have increased apace.

3 The exhausts of road vehicles create a problem of atmospheric pollution in our cities and congested urban areas analogous to the ill-famed Los Angeles.

None of these setbacks need deter those of us in Parliament, in the universities, in industry, trade and commerce, who believe that we must master and order our own destiny and contribute to the future well-being of mankind. The quality of life and improvement of human environment are the greatest challenges of our time. Defeat of pollution and filth is the first and fundamental contribution to such improvement. None of this can be possible without constructive and objective population policies. Because, when considering pollution problems Britain's 226 people per square mile (ten times more than the United States and one and a half times more than India) is gravest of *all*.

Thus, I count myself fortunate to be allowed to contribute this foreword to Dr Ronald Denney's splendid book. As a Member of Parliament, I carry personally a heavy responsibility for all the critically important matters in Dr Denney's book, though I am deeply conscious that to state a problem is not to solve it. But, at least to state is to pose the questions we have to answer; the problems to solve if humanity is to survive.

My duty is to help to evolve formulae whereby improved quality of life and environment may be reconciled with individual freedom and liberty.

Gerald Nabarro, M.P.

House of Commons
1 September 1971

Acknowledgements

Throughout the writing of this book I have been constantly aware of the considerable help and encouragement that I have received from many different people. At no time while I have been collecting data have I met with any obstruction or attempts to prevent me from obtaining information. In fact all the authorities and organisations with whom I have been in contact have supplied me with more than I asked.

I am greatly indebted to Mr F. J. A. Shults, Technical Director of Purle Disposal Services Ltd for spending his time outlining the methods used for industrial waste disposal, and to Mr Higginson, Chief Civil Engineer of the Greater London Council, for a privately conducted tour of the Edmonton waste disposal plant shortly after it was officially opened. I benefited tremendously from the considerable time and help given to me by the officials at the States of Jersey, Bellozanne waste conversion and sewage treatment works. I have been impressed by the interest and encouragement given to me during writing by the many people I have met informally who are involved in pollution control; unfortunately these are too numerous to be mentioned individually.

My two brothers have been very good in supplying articles and advising on their own specialities. I am grateful to my brother-in-law, Edward Barnett, for his excellent and speedy work with my photographs. Once more I must record my gratitude to Miss D. Dickson, F.L.A., deputy chief librarian of the Thames Polytechnic, for the considerable amount of her free time that has been devoted to obtaining journals and publications for me to consult. I have appreciated the wise counsel of Dr R. Marshall, who read my chapter on radiation, and Dr T. Lee, who did the same for my chapter on noise, but am alone responsible for any errors or omissions which have occurred throughout the book.

I am also grateful to the following for permission to reproduce their photographs: Aerofilms Ltd; Aero Service Corporation; Brüel and Kjœr, Copenhagen; Camera Press; The Greater London Council; Keystone Press; The Lancashire County Planning Department; M.S.S. Services Ltd; The National Society for Clean Air; Tom Smith; *The Times*; Topix; The United Kingdom Atomic Energy Authority; United Press International; and John Walmsley.

Finally may I record my appreciation to my very understanding wife, who has tolerated my travels away from home and the long hours necessary to complete this work. I am particularly grateful for the effort she has put in to help complete the manuscript within the schedule that was set.

R. C. D.

Introduction

During the past few years the deterioration of the environment has become a major topic of discussion and controversy. It would, however, be a grave mistake to assume that it is only contemporary generations that have been concerned about the progressive pollution of air and water or the over-stretching of resources. The main difference now, as compared with even a few years ago, is that public opinion is at last beginning to support the enlightened individuals and organisations who in the past have oftened campaigned for improvements without receiving the backing and support that they deserved.

At long last the battles of the conservationists create news worthy of the attention of television and the newspapers. Victories such as those over Stansted airport and Bluebell Wood have been achieved because of well-organised and well-presented opposition. People are beginning to think in national and international terms of cause and effect rather than in a limited parochial context. It is no longer good enough to build a smoke stack that will take the fallout away from one town to the next – but much more important to remove the grit and particles before emission occurs.

Publication of Rachel Carson's best selling book *Silent Spring*, in 1962, added considerable impetus to the concern over the environment. Although the book dealt primarily with the extensive use of chemicals for crop spraying and plant treatment it drew attention to what could occur if man continued to attempt to bludgeon nature into submission with a total disregard for the resulting long-term effects. Needless to say, there are still people who believe that the indiscriminate use of powerful insecticides does no harm either to wildlife or to human beings. These are usually the same people who will advocate pumping untreated sewage into the sea and support the retention of open hearth coal fires irrespective of the obvious pollution problems created.

European Conservation Year, 1970, was concerned with drawing public attention to the overall problems that have arisen from a rapidly increasing, highly industrialised and mainly urban population within the European continent. There were newspaper articles, radio and television programmes, and public displays on all features of pollution, conservation and improvement. In the United Kingdom the whole programme was given tremendous publicity at the beginning by the Reith Lectures on 'Man and Nature' by Frank Fraser Darling in November 1969. As a result, for the first time in the United Kingdom the topic of pollution was on everybody's lips and people were openly discussing such previously taboo subjects as sewage and waste disposal.

The most unfortunate feature of the current concern is that in many cases it has almost reached panic proportions. As a result many genuine problems have been distorted out of size and context. This is not to say that problems do not exist or that they are not serious. The very fact that so little has been done for many years is itself an indication of the backlog of work that must now be dealt with. But it is not enough for writers and television producers to concentrate on drawing attention to what is wrong. They must also show what is being done and where research can work to improve the situation.

Rachel Carson devoted a special chapter in *Silent Spring* to outlining the alternatives to pesticides and toxic chemical sprays. Unfortunately many of the books that have recently been published on pollution and the environment have been almost entirely concerned with telling us all what wicked people we have been to contaminate our planet. At the same time they have almost completely ignored such pioneering work as that of the Royal Commission on Sewage Disposal, which between 1898 and 1910 established the criteria for modern sewage treatment; and they have also forgotten that many flora and fauna protection societies have existed since the nineteenth century.

People who have worked in the realms of air pollution and water purification have, for many years, been given the impression that their researches and achievements have gone unnoticed and unrecorded. This is particularly so in newspaper articles and television programmes foretelling international holocausts in the near future. All too often the impression is presented that we will all suffocate from an excess of sulphur dioxide – if we do not die first from drowning owing to the ice caps melting or are not asphyxiated from the odour of our own human wastes. Television programmes such as *Doomwatch* may be good entertainment resulting from stretching current scientific knowledge beyond the present limits, but the long-term effect is to reinforce the public opinion that scientists are irresponsible and do not care how their discoveries are used or misused. In the long run this over-critical attitude can lead to students being deterred from studying scientific subjects. It can create a situation in which the very scientists needed to control pollution will be in short supply and unable to cope with the severe work load.

Nor is the situation helped by quotations in supposedly responsible publications of meaningless statistics such as 'Finland's inland water is 10 per cent to 15 per cent polluted'. The reader is at least entitled to know if this means 10–15 per cent of all the rivers are totally polluted, or 10–15 per cent are partially polluted, or all the rivers are 10–15 per cent polluted. The misuse of statistics can do more of a disservice to a just cause than almost anything else.

It should be remembered that spoil heaps do not disappear solely from the exercise of vocal chords, nor is the atmosphere any cleaner for all the criticism levied against the internal combustion engine. Any successful steps that are taken to improve the environment are dependent upon people who not only know what needs to be done, but who also know how to do it. There is already enough knowledge and technology to cope with most of our present problems. Enormous improvements have been achieved by people working in the fields of rubbish disposal, smoke control and water purification which have hardly been noticed. What is needed is a better informed public and an acceptance of the fact that every person himself adds to the amount of pollution and waste; and coupled with this must go a willingness to spend money to improve conditions even further.

It will cost money to remove waste tips, to build and modernise sewage works and to clean the air. But the return for this is tremendous: green fields instead of spoil heaps; rivers filled with fish; clean beaches for families to play on; fewer respiratory diseases and better living conditions for everyone to enjoy. This book is concerned with showing how we all contribute to the deterioration of the environment and how the problems which have been created are now being dealt with. At the same time it indicates where attention, research and legislation need to be concentrated for future improvements in the environment to be achieved.

1 The family waste heap

On 5 December 1952 a thick black fog descended upon London. When it lifted four days later the word 'smog' had become part of the English language, and as a result of that suffocating blanket about 4,000 people died. The only new feature of that particular combination of smoke, acid and poison was that it lasted longer than any previous London fogs of that type had done. But there was nothing different in the cause or composition of what had become an accepted part of our, so called, civilised society.

As far back as 1661 John Evelyn had tendered advice to King Charles II on how to clean up London's smoky, foggy atmosphere. But it was not until nearly three hundred years later that the first major legislation to restore clean air to Britain's industrial areas was introduced. The Clean Air Act, 1956 followed from the recommendations of the Beaver Committee that had been set up after the 1952 smog, and started the process of improving the atmosphere which has since been continued by the Clean Air Act, 1968.

The occurrence of the London smog and its results was probably the first catastrophe to bring home to people the extent to which the environment was deteriorating. At the same time the effects of the Clean Air Acts have been remote from the majority of the population. For the greater part it has only meant changing to smokeless fuel and changing the grate in the fireplace. In many areas any decrease in smoke pollution has been due more to slum clearance and the construction of centrally heated tower blocks rather than a deliberate policy to improve the atmosphere.

Despite the frequent occurrence of fogs and smogs it is fair to say that until the 1950s the general public was for the great part blissfully unaware of the deterioration of the environment and the great industrialised nations of the world tended to think of their pollution and environmental problems in isolation from each other. Poisoned rivers, atmospheric fallout and rubbish heaps were considered necessities which accompanied full employment and higher standards of living.

Even today the average person has little concept of his own personal contribution to the deterioration of the environment or the part that he should be playing in preventing the human pigsty from sinking in its own filth. The dirt falling on the washing in the garden is always from some other chimney. The froth on the river is due to other people using detergents. The waste paper blowing about the streets is always dropped by visitors from other places; and the noisy car or motor cycle spluttering and back firing along the road belongs to some reckless teenager, of course.

Pollution is a personal thing – we all produce our own share of waste products and are all irresponsible in the manner in which we misuse and maltreat the environment. It is true that we are not entirely to blame for such things as dirty exhaust gases or oil on the beaches, but every day we all add substantially to the wastes of the community.

Just consider an average family living in their suburban semi-detached house. About the only time they can be considered to be non-polluters is while they are asleep. The rest of the day, from the time they wake up until the time they go back to sleep, they make their own personal contributions to sewage, household waste, water pollution and air pollution.

Before Mr Everyman wakes up in the morning his central heating may have automatically switched itself on. Unless it is electric it will run on gas, oil or solid fuel; and all of these not only produce carbon dioxide and water as the normal products of combustion, but also form a small amount of an obnoxious gas known as sulphur dioxide. The large amounts of carbon dioxide produced are eliminated from the atmosphere by photosynthesis in plants which use it and return free oxygen to the air in its place. There is a possible long-term danger here in that if the carbon dioxide level rises too high the plants will be incapable of converting it all back to oxygen. If this situation arises it may result in global temperatures increasing because of the glasshouse effect for which carbon dioxide is responsible.* As the carbon dioxide level increases, more of the sun's rays are trapped and temperatures rise. The situation will be aggravated even more if Mr Everyman later in the day decides to reduce the number of photosynthesising plants by uprooting the bushes and cutting down the trees in his garden.

The sulphur dioxide, also produced during the central heating combustion, arises from the small proportion of sulphur compounds found in all fossil fuels. As sulphur dioxide is a gas, it mixes with the normal gases in the air and can travel vast distances. Eventually it comes back to earth by being absorbed by plants either as the gas or in rain as sulphurous and sulphuric acids. In Sweden it has been claimed that sulphur dioxide originating in Great Britain and Germany is acidifying surface waters and stunting the growth of the pine forests.

Having woken up to a centrally heated house, Mr Everyman and his family prepare for the day with their visits to the lavatory, washing and shaving and so start the first of their daily contributions to the sewage works. All their water-borne wastes, consisting of urine, faeces, paper, soapy wastes and anything else thrown down the lavatory or sink will start on a journey lasting several hours that in a well-regulated district will result in their eventual removal before the water is passed back often into the river from which it was originally taken.

If Mrs Everyman is really up-to-date she will also dispose of a large quantity of her solid kitchen waste by feeding it into a kitchen sink disposal unit in which it will be macerated and flushed into the liquid waste pipe to join the rest of the sewage. Alternatively, her kitchen waste will be placed initially into a plastic bin liner in a pedal bin and later removed to the dustbin or paper refuse sack supplied by the local authority. The accumulated household rubbish will be removed later in the

* This is dealt with in more detail in Chapter Six.

week by the dustmen and taken to the refuse disposal works where it may be sorted, macerated, used for composting or dumped.

The waste from the Everyman household consists of a tremendous mixture of materials. It includes bottles, tin cans, plastic containers and bags, food residues, dust and dirt, newspapers, rags and sometimes garden waste. Each component presents its own problem for disposal and incorrect or inadequate treatment at the disposal plant can create further pollution problems.

While Mrs Everyman is beginning her daily chores her husband will be on his way to work and her children on their way to school. If her husband uses his own transport he will have started his car by pulling out the choke to enrich the petrol supply to the engine. Before ignition occurs, petrol vapour will come out of the exhaust, and after ignition, while the choke is still out, the engine will produce a mixture of gases including carbon monoxide and water, with some nitrogen oxides in addition to a proportion of unburnt petrol. If the petrol contains an organo-lead anti-knock compound, then a small amount of lead will also come from the exhaust. When the engine of the car has warmed up and the choke been pushed in the amount of unburnt petrol will be considerably reduced, but the other emissions will continue while the engine is running.

The Everyman children will probably walk to school or travel by bus. In the latter case the vehicle will have a diesel engine. This is notorious for its production of black, oily smoke which contains unburnt hydrocarbons and nitrogen oxides as well as carbon monoxide and carbon dioxide. Under normal running conditions the engine of the bus should not emit very much in the form of visible pollution but black smoke is produced when incomplete combustion of the diesel fuel is taking place in the engine. This is usually the result of extra fuel being injected into the engine when under load travelling up an incline.

While her husband is at work and her children at school Mrs Everyman will carry out her housework. For her washing up she will probably use a liquid detergent that will be discarded down the sink in the washing-up water. Later she will use a powder detergent in her washing machine and again add to the problems of the sewage works when the dirty water is spun from the clothes. During the past twenty years sewage works have found it very difficult to cope with the ever-increasing quantities of detergents arriving in domestic liquid waste, and often sewage foaming becomes a serious problem as some detergents are difficult to remove despite modern purification processes. On occasions foam develops to such an extent that it gets carried away by the wind and may finish up in the back garden of the house from which the original waste came.

If Mr Everyman works in an office he will spend most of his day at a desk reading, writing and making telephone calls. As the day proceeds his waste-paper basket will fill up and at the end of the day it will be cleared out and its contents added to the pile of rubbish to go to the refuse depot. By smoking a cigarette he will pollute the air in his office with many different chemical compounds which are the constituents of the tobacco tars he exhales. Some of these are dangerous and can lead to lung cancer if frequently inhaled. The used matches, cigarette ash and cigarette ends will all eventually add to the pile of waste cleared by the office cleaners at the end of the day. The whole lot will be passed for treatment to the refuse depot, along with the plastic coffee cups from the automatic vending machine, the paper towels from the wash rooms, and the used ball pens discarded during the day.

Mr Everyman's children will also have done their share during their day's labours to add to the pollution worries of the world. Their efforts churned out on paper will eventually finish up in the waste bin and go with the refuse collection. If they have done any practical chemistry or biology in the laboratories there will have been small amounts of chemical wastes in the form of acids, alkalis, or solvents washed down the sink to pass to the sewage works. Their empty milk bottles will have to be washed at the milk depot and if they dislike the school dinners they are served, the uneaten food will be taken away and treated to be fed to pigs or used in compost for agriculture.

Even the relatively innocent occupation of shopping that

Mrs Everyman has to do most days adds to the problems of waste disposal. The cans and packages that she purchases in the shops and supermarkets are transferred to the shelves from large cardboard containers which have to be folded and taken away for disposal. At the check-out she will be given a list of the prices she has paid. This may finish up on the floor of the store or be later discarded in the street to add to the litter of the town. Later on all the wrappers, plastic bags, boxes and cans containing her food will be thrown away and pass into the dustbin for disposal at the refuse depot.

Even when the family has returned home and eaten its evening meal its creation of waste and pollution has not finished for the day. If Mr Everyman has a garden he will have to attend to it regularly to keep it respectable. This will mean periodic weeding of the flower beds, regular cutting of the grass and occasional pruning of the bushes and fruit trees. If he is a keen gardener he will dispose of his clippings by building a compost pile that will later be fed back into the earth. It is more likely, however, that he will take other steps to get rid of the waste. The easiest method will be to have a bonfire at the end of the garden. The green, often damp, garden residues will be left to smoulder for several hours producing smoke which will blow away from the Everyman's home, possibly into somebody else's back window.

The fault may not be entirely Mr Everyman's if local facilities for the disposal of garden waste are virtually non-existent. An alternative disposal method may be forced upon him if neighbours object to their washing being covered with black specks. Under these circumstances he is likely to load the rubbish into plastic or paper sacks and drive into the surrounding countryside to find a nice isolated wood or ditch into which the sacks can be surreptitiously dumped. The rubbish is then, for him, out of sight and out of mind, being transferred to some farmer who will have to get rid of it properly when he finds why a particular ditch is failing to drain correctly. If Mr Everyman is fortunate enough in having a local authority that either agrees to collect garden waste or alternatively supplies a local open dump his problems are solved for him. He may still, however, like to see the plume of smoke rising above the incinerator if only to advertise the fact that he has done some gardening.

It would be nice to think that this is the end of the Everyman potential to contribute to the desecration of the environment, but even more is possible when they are relaxing or on vacation. Even a simple thing like an aeroplane flight taken by the family to its holiday resort results in air pollution from the jet engines as well as high noise levels for inhabitants near the airport. In addition to this the Everymans will be served meals in plastic and aluminium wrappers that will need to be disposed of at the end of the flight, along with the sewage from the cabin toilets.

A day trip in the car to the seaside will automatically contribute to the air pollution, and this may become substantial if the vehicle is idling in traffic jams for long periods. During the journey the family will eat sweets and wipe their sticky mouths with paper handkerchiefs. If they are particularly thoughtless the wrappers and tissues will be discarded through the car window to lie in the kerbside until cleared up by a roadsweeper. If Mr Everyman is a smoker his matches and ash will similarly be flicked out of the window. If he is very careless his glowing cigarette-end will finally be discarded in the same way, and may set fire to the grass verge or even cause an accident by flying into the face of another road-user.

On the beach the Everyman children may turn on their transistor radio for the morning pop concert. In the belief that everyone should share their pleasures they may turn up the sound as loudly as possible with the result that the sought after relaxation and quietness is ruined for the majority of the other beach users. The Everyman family will then spend the day swimming, sunbathing and eating. Having brought sandwiches and fruit from home they will lie on the beach and unwrap their packages. If it is a nice, lazy day, it may be too great a trouble to walk to the waste-paper bins twenty yards away. So the grease-proof paper and the plastic bags will whisk along the sands in the wind. Eventually they may finish up being carried out to sea or be cleared up by the beach-cleaning patrol. Before going home, Mr Everyman will take his beach tray back to the kiosk, and on the way he may drop one of the lemonade bottles causing it to fall on a stone and smash to pieces. But that is not likely to worry him beyond the fact that he has lost some of his deposit. His family will be leaving the beach in a few minutes and 'after all, the tide will come in and cover the glass over in a short time'. So the glass adds a more dangerous aspect to the pollution and litter on the beach, and will probably lead to someone else suffering at a later date.

The Everyman family would probably act in a similar manner if they went for a drive in the country or visited a football match. They would leave behind them a trail of waste and pollution to be cleared up by other people. Their excuses would be that it is not their fault that plastic bags do not rot away; they are not responsible for the shortage of waste-paper bins and other people can also switch on their portable radios if they wish to.

This particular family has been given all the least desirable habits and attitudes imaginable. Unfortunately such characteristics are only too common.

Pollution and environmental problems occur not only from accepted and established modes of conduct such as driving motor vehicles or using the lavatory but are aggravated by the indiscriminate disposal of rubbish and lack of use of facilities that are available. It is difficult enough to cope with wastes and pollutants that have been produced by acceptable living, but the problem is greatly magnified when even a moderate code of conduct is not followed. In any case, is it such a great hardship to take rubbish home if there is no litter bin around? People who throw their waste from cars or who dump their old bed mattresses in the woods are as much to be condemned as the Didikois leaving stripped car bodies scattered along the roadside. Such problems may have arisen, at least in part, because of the planned obsolescence and excessive packaging of our civilised society; a subject admirably analysed by Vance Packard in *The Waste Makers*. But this is no excuse for the grossly irresponsible attitude assumed by a large number of people who assume that responsibility for restoring the quality of the environment rests with someone else.

The problem has ceased to be the simple one of smoke from one garden blowing into another, or whether or not a particular town will build a sewage works. Pollution of a river by one town can deny its use for drinking water to another. Oil discharged at sea by the tankers of one nation can pollute the beaches of half a dozen other nations. The indiscriminate use of toxic chemicals by one farmer can destroy the wildlife that gives pleasure to the many, and is in any case part of the natural order of things and as such to be protected as far as possible.

As a nation becomes wealthier and enjoys a higher standard of living so the amount of waste it creates also increases. In the U.S.A., refuse has now reached the incredible figure of 1,000 lb per head each year. The comparable figure for the United Kingdom is about two-thirds of this at 650 lb per head, but is increasing rapidly. Because of this level of waste many refuse disposal plants are, or soon will be, incapable of coping with the loads imposed upon them. This remark applies equally to sewage treatment, and thus the main source of pollution in many rivers is untreated or only partially treated sewage.

Man is probably the dirtiest animal in the world. He pollutes and contaminates almost everything with which he comes into contact. His discarded rubbish and pollutants can be found in even the remotest parts of the world, from the frozen lands of the Antarctic to the hot deserts of the Sahara and even in outer space. The big question now is whether or not he is prepared to pay heed to the dangers which have been so vividly drawn by different experts. Have we matured enough now to use our knowledge and technology not only to raise the standard of living, but also to reverse this other downward trend which could lead to self-destruction? The decision rests with every individual as well as with every nation.

2 The dark satanic mills

The standard of living of the people in any country depends upon their industry, whether it be agricultural or in urban factories. Because the United Kingdom was the first nation to experience an industrial revolution it was also the first to suffer from the problems that arose from a combination of rapidly expanding towns, inadequate housing, and lack of industrial planning and control. Unfortunately, many of these problems, created from ignorance in the last century, have still not been fully corrected in this. Other nations followed the United Kingdom in making the same mistakes, which allowed industrial progress to occur without adequate consideration for the effects on human living conditions and the general environment. In recent years new industrial methods have created further waste and pollution problems which all too often are ignored or not fully appreciated until some accident occurs.

Frequently industries are responsible for releasing substances as waste which would more than cover the cost of recovery if they were trapped and collected. It is a fallacy to assume that this attitude only arises from the profit motive, for state-owned concerns are certainly no better than private companies. Industry is the main cause of water pollution; it is responsible for 30 per cent of air pollution (excluding the contribution from its motor vehicles); it is almost completely responsible for radiation waste; it is the greatest spoiler of the land and also produces most of the undesirable noise.

What is surprising is that it is not fully appreciated that the very existence of waste from any manufacturing process is evidence either of inefficiency or of failure to make use of by-products. Thus power stations failing to wash their smoke emissions are wasting thousands of tons of sulphur in the form of sulphur dioxide, chemical companies polluting rivers are losing expensive acids and solvents, motor vehicles and aeroplane engines creating unnecessary noise are wasting power, and the indiscriminate dumping of rubbish is a waste both of useful material and of good land. Unfortunately it all happens so easily, and all too frequently expanding companies fail to increase the capacity of any treatment plant originally installed years before.

Let us look at an hypothetical chemical company which refuses to comply with even the limited regulations that already exist. Apart from the sewage from the toilets provided by the company for its employees, there is a great deal of chemically polluted water-borne waste. Any chemical spillage, which might be acidic or toxic, will be washed into the gullies in the production shops and led into the main sewage system. The same situation will occur with washings from the dirty equipment after reactions have been carried out.

If there is direct drainage from the factory to the main sewage system not only will any rain water from the factory site drain straight to the main flow, but any major spillage from an acid tanker or solvent from a burst drum will probably reach the sewage works in a highly concentrated form and completely disrupt the water purification processes. In cases in which the sewage water is discharged into a river it is possible for many chemicals to pass through the sewage treatment without being removed, and thus contaminate and possibly poison the river for other users.

The chemical factory will also use large quantities of cooling water, probably from the river, in giant condensers to cool down refluxing chemical reactions. During this process the water itself becomes heated and loses its oxygen. If the warm deoxygenated water is returned to the river it will be incapable of supporting either fish or plant life and will help to kill the life in the river. The problem will be greatly increased if the company is careless enough to discharge any of its waste chemicals directly into the river. In this case even small concentrations of toxic chemicals could have a devastating effect on the natural state of the river.

Our non-thinking chemical company still has problems in its disposal of solid chemical residues. It may decide to try and wash these down the drain gradually. But if they are relatively insoluble in water this can take a great deal of time, so the alternative is likely to be a dump. The dumping may take several different forms, each with its own attendant disadvantages if carried out on a thoughtless basis. The waste might be dumped into a river late at night in the hope that the flow will wash it away before daylight. More commonly, it will be dumped on a waste site in the factory grounds. This, however, may lead eventually to its entering an underground stream by dissolving in rain water by which it may be carried to a reservoir and pass into drinking water. In addition to this the indiscriminate dumping of one chemical waste on top of another can lead to fires and even to explosions. An alternative dumping technique that is sometimes used is to discharge it on any convenient space of land in the countryside with the attendant risk that a hiker or an adventurous child may stumble into it and possibly suffer from chemical burns or poisoning.

In addition to this the company will almost certainly have its own power house for generating electricity and heating water, and no doubt its lack of control will extend to adding to the smoke pollution in the area.

The reader may feel that this is all stretching the imagination a little too far. Whilst it is unlikely that any one company would successfully carry out such large-scale polluting for very long, it is correct to say that each one of the inconsiderate waste disposal methods mentioned has been used on innumerable occasions. There are many companies that are well aware of the nature and quantity of the wastes they are discharging into the atmosphere, the rivers and on the land and yet do very little to ensure that the discharges are harmless. It is only necessary to look at some of the more obvious cases of pollution to appreciate fully how serious this lack of control and public conscience is.

In North America the Great Lakes cover an area of nearly one hundred thousand square miles and drain a land area of about that size. The shallowest lake of the five is Lake Erie, and it is this which has become the most heavily polluted of all the lakes. Obvious evidence of this has been seen from the rapid decline in the numbers of fish caught in the lake during recent years. This has come about because the discharge of waste into the lake has grown so great that the lower depths of the water are almost completely devoid of dissolved oxygen. Around the shores of Lake Erie are five big cities, including Cleveland and Detroit. It is from the industry and population in these areas that the pollution originates. The Detroit River runs into the lake and studies have shown that it is adding a continuous quantity of oil, organic chemicals, inorganic phosphates and chlorides, as well as bacteria and undissolved solid materials. The natural processes within Lake Erie are incapable of coping with such a deluge, and it has reached the stage at which the settling of solids and the decomposition of chemicals and wastes have killed most of the life in the lake. It is, in fact, so heavily polluted that a serious suggestion was made that the only way to clean it was to drain it, dredge it and refill it.

An even worse state of pollution exists with the Cuyahoga River in the State of Ohio. This is the only stretch of water in the world known to have burst spontaneously into flames. The river is effectively an open industrial sewer carrying large quantities of chemicals and oil waste. Whenever the right combination of substances occurs, the river becomes a fire hazard (and contributes to the air pollution at the same time).

It would be a mistake to believe that such situations do not exist in other countries. In fact it is possible to find similar instances of pollution in practically every major industrialised nation. It is only necessary to look at the River Irwell in England to get some idea of the blind eye that so much of industry has turned to pollution. The Irwell starts off its life as a clean Pennine stream. It eventually becomes the Manchester Ship Canal and finally joins the Mersey before passing out to sea. During its journey it has discharged into it the effluents from several paper mills, two dye works, sundry sewage works, numerous factories and several power stations. The originally clean river completes its journey as a dirty, lifeless, deoxygenated flow referred to by one writer as 'a festering sore'.

Many rivers in Great Britain are in a similar state, those in South Wales being particularly bad. It has been reported in *The Ecologist* that the River Ebbw possesses virtually no animal or plant life after it has been polluted by the Richard, Thomas and Baldwin steelworks. Eighty tons of solid waste are discharged into the river every day, coating the bed with a thick layer of iron hydroxide. There is a more pleasant side to this story, however, as this firm are in the process of constructing a plant to carry out effluent treatment in an effort to reduce the pollution to about 3 per cent of its current level.

Practically every form of industrial process produces its own type of air pollution, which may range from being simply an unpleasant smell causing no harm to being a dangerous invisible gas, such as hydrogen sulphide or sulphur dioxide, or undesirable airborne particles which settle everywhere.

Industrial contribution to air pollution usually arises from smoke stack discharges. The traditional idea is that the higher the stack the larger the area over which the solid particles will settle, and the greater the height at which the gaseous constituents will become diluted by the atmosphere. It should be remembered that the smoke stack does absolutely nothing to decrease the total amount of discharge. People who succeed in getting a smoke stack extended in height because its fall-out settles on their washing make no contribution to solving the pollution problem. All they do is drive it away to bother someone else. The real answer is to tackle the source of the trouble.

Whilst the emissions from brickworks or cement works are mainly particles of dust, those from steel works and chemical plants are predominantly gaseous. Failure to remove these emissions leads at least to dirty washing, spoilt paintwork and begrimed buildings; or, at worst, to rotting brickwork, stunted vegetation and serious chest and lung diseases.

It is difficult to grasp fully the extent to which industrial growth has led to the despoliation of so much that was once beautiful wooded or pasture land. It is only when a tragedy such as that at Aberfan occurs that public attention, for a short space of time, is focused upon rubbish tips. Over the centuries man has excavated, sifted, selected and rejected, and the results of his efforts cover the face of the earth like pock marks and boils. Side by side with the holes in the ground leading to the coal mines are the spoil heaps of shale and slate. Giant excavators have gouged out great chunks of the earth to produce the raw material for our concrete empires and have left us with misshapen pits filled with water and a despoiled countryside. The railways have desecrated vast areas in the country, as well as in the towns and cities, with rusting, overgrown sidings and unused land. Brickworks, china-clay companies and steel works have all left their rejected materials in giant pyramids reaching to the sky.

Official government figures claim that only about 130,000 acres of Great Britain can be classified as derelict. This does not, however, include land that is considered to be in active use, and, as John Barr has pointed out in *Derelict Britain,* the official figures excluded such tips as that at Aberfan which was still being used. Other estimates have put the area of derelict land as high as 250,000 acres. In addition it is claimed that for every derelict area of land there are ten acres that can be classified as sub-standard. At present the rate of reclamation of the derelict areas is hardly keeping pace with the rate at which other land is degenerating to this level.

This is not to say that nobody is doing anything about it. The most active reclamation is proceeding in Wales, where people are now less prepared to live with those hideous legacies of previous working generations on their doorsteps. The cost of reclamation is about £2,000 per acre but the return for this is new playing fields, landscaped hills, better parks, more trees and better scenery.

Radioactive waste disposal problems have arisen during the past twenty-five years, and the amount seems destined to increase fairly considerably as more and more nuclear power stations are constructed. Disposal of these wastes cannot be achieved simply by chemical methods as this has no effect on the concentrations of the actual radioactive elements present. Fortunately, the amount of radioactivity automatically decreases at a constant mathematical rate, which differs for each isotope present, and over a period of time there is a reduction in the emitted rays that are harmful to living organisms including Man. Whilst there is a minute amount of radioactivity in everything it is undesirable that any significant increase in this level should occur because of careless disposal of the radioactive waste materials. Disposal in the sea has been particularly popular because of its vast capacity. This has led to an attitude of 'out of sight out of mind', which has tended to ignore the fact that winds and currents can lead to the wastes being carried back into estuaries and to beaches. Another danger is that the radioactive compounds may be absorbed by fish and finish up on a plate being served for dinner. So far the long-term effects on marine life in general have only been poorly established.

Disposal of radioactive waste in the ground suffers from similar shortcomings in that there is always a danger that it will be leached from its burial site and pass into rivers or streams and eventually into the domestic water supply. In addition to these forms of radioactive wastes there is a continuous discharge of small quantities of radioactivity into the atmosphere from factories and laboratories in which gaseous radioactive materials are used. Emission of this type can produce a radiation hazard directly from the radioactive cloud; it may alternatively be inhaled and cause damage in this way or it can be ingested by an animal and finish up being eaten. It has been found that fish will selectively concentrate the isotopes of strontium, caesium and phosphorus in their tissues so that the results of even substantial dilution of radioactive wastes can be reversed and lead to unwelcome and dangerous levels of radiation.

One form of industrial pollution that is giving rise to worry on a considerable scale is that of oil discharges, whether accidental or deliberate, into the sea. The situation has become more acute in recent years with the development of supertankers and the general increase in the amount of oil being shipped around the world. The *Torrey Canyon* disaster led to the first real appreciation of the havoc that would be caused by substantial quantities of oil. Since then other oil tankers have been involved in accidents and spilt their obnoxious cargoes, killing off marine life and ruining the beaches.

In the U.S.A. drilling for oil at Santa Barbara met with more success than was anticipated when the oil under pressure escaped into an underwater fault and within a short time erupted in the sea from five different points. Another recent accident was the collision of the Spanish ship *Monte Ulia* with the refinery jetty at Coryton in Essex. This accident alone resulted in the release of five hundred gallons of crude oil into the River Thames and the pollution of many miles of the coastline.

It is not only accidental spillages that have contributed to the pollution of the oceans. Reports are continuously being made of ships discharging oil waste at sea contrary to international agreement. This was particularly highlighted by Thor Heyerdahl after his trip on the papyrus raft *Ra*, when he said that during the fifty-two-day voyage they drifted for six days in mid-ocean in a continuous area of polluted sea. It is quite obvious that control and supervision in this realm is grossly inadequate. Not only is more research necessary to deal with the removal of oil discharges, but the problem must be dealt with more severely at source.

Another area requiring more public attention is noise. Unfortunately what is unpleasant noise to one person can be the most desirable sound to another. Also there is too great a tendency for the noise-creators to adopt the attitude that noise must be tolerated if it means that aeroplanes can travel faster than sound. There is little point in exchanging the air pollution of steam engines for the noise and nuisance of diesel engines as the net gain to the environment is zero. At present the majority of the population is being subjected to excessive noise because of the wishes and desires of the minority.

Whilst millions of pounds and dollars are being spent to develop new engines for cars and aeroplanes only a small proportion of this is devoted to research to reduce the amount of noise these engines produce. It has become quite obvious that laws and bye-laws are now inadequate to control effectively the continuous bombardment on our ears. The quietness of our towns and villages is being shattered by giant lorries grinding their way through narrow streets. The countryside is resounding to the deep-throated reverberations of sports cars out on rallies, and large areas of all countries suffer day and night from jet aircraft going in to land at major airports. It has been demonstrated quite clearly that over a period of time excessive noise can not only damage a person's hearing but also create physiological changes in people in the form of decreased blood circulation and muscular tension. Undoubtedly the major concerns of people living in noisy environments are loss of sleep, interruption of conversations and general nervous stress. At present the situation is deteriorating because of the increases in flying and motoring and also the cheapness of portable radios. Because of this not only is it necessary for more work to be done on methods for reducing noise levels but people need to be more concerned about steps they can take themselves to prevent sound-penetration in their own homes.

Agriculture is still the largest single industry in Great Britain, and whilst it does not use smoke stacks and rarely discharges its wastes into rivers deliberately it does in its own way add to the pollution of the country. The farmer uses chemicals for two main purposes: to fertilise the ground to improve the yield and quality of his crops, and to treat plants in order to remove unwanted insects and parasites. For the former purpose the main active materials will be inorganic phosphates and nitrates. Some of these eventually finish up in the rivers owing either to erosion or to leaching and they create their own particular problems. Phosphates act as important nutrients for algae in the river and promote the growth of these and other plants which use up the oxygen in the water making it unsuitable for fish. If nitrates reach the domestic water supply they can cause illness in both animals and human beings. This is because they break down to nitrites, which are the cause of methemoglobinemia in newborn children (blue babies). It is obvious that there is a great need for care in this matter, particularly in the case of farms where drinking water is drawn from a well, as this could have a serious effect on any young children in the vicinity.

The word 'pesticide' is an inclusive expression which covers a wide range of different chemical compounds employed for a variety of purposes. It embraces insecticides, herbicides, fungicides, rodenticides and fumigants. There is nothing new in the use of such compounds. Paraffin was used many years ago for treating ponds to keep down mosquitoes, lead arsenate has been extensively employed for controlling a variety of garden pests and the extraction of nicotine from tobacco for application to fruit trees was first carried out over two hundred years ago. What has changed is the development of the wide range of chemicals that are now available for eliminating practically every form of pest likely to interfere with the successful growth of a crop. In the process a situation has been created in which not only have many non-harmful animals been almost completely exterminated in various parts of the world, but the very pests they were aimed against have often developed an immunity. At the same time the chemicals employed are entering our rivers and becoming incorporated into our bodies. Public awareness of the danger has arisen from the tremendous publicity given to D.D.T. and its related chemicals which persist in body fat for very long periods. However, the very focusing of attention on one particular group of chemicals has also tended to distract attention from other misapplications.

Industry has a key part to play in improving the quality of our life by taking the steps necessary to prevent pollution of air and water and to repair the ravages of centuries of desecration of our national heritage. The time has now come for us to take drastic action to stop the rot and to repair the damage so that posterity will look back on this generation and say 'They cared about the future'. A great deal of pioneering work has already been carried out in this field. Many of the answers to our problems already exist and further research is already under way. What is required above all is a greater dissemination of the available knowledge and the support of public opinion for spending the money so essential to improve the environment. The world's pollution problems will not be solved overnight, nor will they be solved cheaply, but it is vital that we make a start now.

3 The struggle for clean water

Although two-thirds of the Earth's surface is covered by water, most of it is in the form of seas and oceans and contains so many dissolved salts that it is unsuitable for drinking or industrial purposes without some form of treatment. In a highly industrialised society the demand for water is tremendous and increases every year. Domestic use alone in Great Britain requires fifty gallons of water per head of population every day for washing, cooking, sewage and drinking. Industrial demands on the water supply are even more gargantuan, with enormous quantities being required for most manufacturing processes. Industry uses water for electricity generation, heating, cooling, water power and for incorporation into products. It has been calculated that the manufacture of one ton of steel requires 50,000 gallons of water, and the brewing of one pint of beer needs 1 gallon of water. The largest industrial users of water are iron and steel manufacturing, the chemical industry, oil refineries, paper mills and power stations. It is only fair to point out that, unlike the domestic water user who can use almost as much water as he wants for a very small annual charge, the industrialist pays for his water on a metered basis. Because of this he is automatically encouraged to re-cycle and re-use the water for which he is being charged.

Because of the large requirements of water, often of a particular quality, industry originally developed during the industrial revolution in places where there was a cheap and easily available supply of water. This was the cause of the localised development of the cotton industry in Lancashire and for the quality of the national drink in Scotland. Very few artesian wells are used now, so that water requirements are satisfied by direct extraction from rivers and estuaries or from reservoirs. Most of the water is eventually returned to the rivers, either by direct discharge from the manufacturing plant or through the sewage works. In the majority of cases it is less wholesome after use than it was before and because of this pollutes the river to which it is returned.

It is entirely wrong to talk in terms of 'pure' water for domestic users. Pure water, by definition, means water that contains absolutely no impurities at all. The nearest that most domestic users come to ever encountering pure water is when they put distilled or deionised water into their car battery or into their steam iron. In fact, pure water is completely tasteless and very unpalatable. Most people would consider it to be unsuitable for normal use and would almost certainly object to their local water board about the poor quality of the water if they were supplied with it. It is more correct to talk in terms of 'wholesome' water, which for drinking purposes should be colourless, odourless and free from suspended matter. In addition to this it should not contain any bacteria or dissolved substances likely to be harmful to the consumer. Producing water that meets these standards requires a high degree of treatment and control, and it says much for the water companies in the United Kingdom that there are so few complaints about their supply. The demand is, however, now beginning to exceed the supply, and if the pollution of our major rivers is not substantially reduced during the next few years there will be serious water shortages before 1980.

Initially all the water in our rivers comes from rain or one of its frozen modifications. On its way through the atmosphere it dissolves such gases as carbon dioxide and sulphur dioxide. Percolation through the soil to the rivers and lakes leads to further solution of salts and organic material in addition to the carriage of undissolved solids. Also the natural characteristics of the water can be substantially changed by man's agricultural and manufacturing activities. Before the authorities can release water for domestic use they must ensure that any unpleasant or injurious contaminants have been removed.

This necessity to ensure that the supply is potable means that numerous continuous chemical and biochemical analyses have to be carried out to establish the levels of sulphates, nitrates, ammonia, dissolved oxygen, metals, acids, organic substances and certain pathogenic, disease-causing bacteria in addition to any special substances that may have been released into the water source in a particular area. The presence of any substance producing an unpleasant taste or colour to the water supply necessitates special purification processes being applied to it. For drinking purposes purification methods have reached such a high level of efficiency that heavily polluted river water and sewage effluent can be treated well enough to produce a wholesome water supply.

In the purification process reservoirs are initially used to store water for about two to three weeks before it is passed to other purification stages. The action of sunlight and atmospheric oxygen on the water lead to a significant reduction in the bacterial content of the water as a first step in the purification. This does not occur with water taken straight from the rivers and passed to the further stages of purification without being stored. Water from the reservoir or river is screened to remove suspended matter before it passes to coagulators in which it is treated with aluminium sulphate and activated silica to remove colour, turbidity and any remaining suspended matter. The coagulated material settles out in sedimentation tanks and the water is passed on for sand filtration. Passage through sand filters is a well-established procedure for removing suspended matter after the coagulation process. This process is now commonly carried out under pressure to enable large volumes of water to be processed within a relatively short space of time.

At this stage the water still contains bacteria which could be harmful, and it is necessary for sterilisation to be carried out. A large number of chemicals have been used for this purpose, most of them containing chlorine in some form or another. Chlorine gas was first added to the domestic water supply on an exploratory basis in 1897 and its success was quickly followed by the use of sodium hypochlorite, bleaching powder and later by chloramine, all of which act as bactericides. The first full-scale plant for the chlorination of water was installed at Reading in 1910 and this led to its adoption as a standard purification method throughout the world. Any residual chlorine imparting a taste to the water is removed by the addition of sodium thiosulphate or sulphur dioxide. In some instances sterilisation of water supplies has been achieved by irradiation with ultraviolet light. Its use, however, is limited to clear water sources as particles and dust absorb the rays and reduce their efficacy.

Following sterilisation the water is often aerated by the injection of air under pressure as it passes through the purification plant. This process helps to remove any residual odours and leads to the precipitation of any dissolved iron. The general taste of the water is improved by the inclusion of this step in the overall purification.

There are additional purification stages that can be introduced for the removal of particularly persistent impurities where necessary. These include treating the water with specially activated carbon to remove odours and colours. It has also been recommended for the removal of any radioactive elements that may have become dissolved in the water. In cases in which the water has a high degree of hardness because of undissolved calcium and magnesium salts, special softening processes may be introduced either by treating with lime-soda, or more commonly these days by the use of ion-exchange units in which the calcium and magnesium salts are completely removed.

Water treatment plants are constructed to cope with a particular volume of water designed to satisfy the needs of the area they are intended to supply. Under normal circumstances the quality of domestic water produced should satisfy both the Ministry of Health bacteriological standards and the normal requirements for a wholesome water supply. Difficulties occur when the demand for water reaches the capacity of the plant, as any attempt to operate the plant beyond its design limits results in a lower degree of efficiency at each purification step and a poorer quality of water being produced. This has become a major problem in some parts of the country with the rapid increases in population arising from developments of new housing estates. The situation can be further aggravated if there is any significant deterioration in the quality of the impure water passed into the purification plant. On occasions water authorities have had to stop extracting water from rivers when the level of pollution has risen owing to the discharge of some obnoxious chemical higher upstream.

As the populations of countries increase it will be necessary to construct more reservoirs and water purification plants to satisfy requirements. Methods for purifying sea-water to make it suitable for drinking have been under investigation for a long time, and plants for this purpose have been in use in some countries for many years. Part of the water supply on the island of Guernsey is obtained by flash evaporation of sea-water, and this process has been found to be economically competitive with conventional methods for purifying water. Sea-water can also be made potable by freezing out the impurities and brackish water purified by a process known as electrodialysis, in which the impurities are removed by attracting them to electrically charged ion-exchange membranes through which the water passes.

These methods for increasing water supplies are under active consideration in hot countries such as Israel, in which the water demand is likely to exceed the supply within the next few years. Two demonstration plants for electrodialysis have already been erected in southern Israel as the first steps to a full evaluation of the process. Desalination by electrodialysis is feasible for brackish water containing up to ten thousand parts per million (1 per cent) of sodium chloride and has been used in South Africa for producing water at the rate of 100,000 gallons per hour containing only five hundred parts per million of sodium chloride.

The best contribution to an improved water supply in most industrialised countries would be by a decrease in river pollution; at present many rivers are so heavily contaminated that they are completely unsuitable for any use to be made of them for drinking purposes. In the U.S.S.R. it has been reported that nearly a quarter of a million miles of rivers are polluted. Even in the United Kingdom there are five thousand miles of polluted rivers, many unable to support any form of aquatic life. Typical of the prevailing state of affairs in many parts of the United Kingdom is the River Tyne, which has forty million gallons of untreated sewage discarded into it every day by the Tyneside conurbation. Over a period of one hundred years the Tyne has degenerated from an excellent salmon-fishing river to an open sewer almost entirely devoid of fish. Fortunately the local authorities in this particular area are now co-operating in the development of a comprehensive sewage disposal system costing thirty-five million pounds and intended to start operation in 1977. To bring this about has meant that twenty different local authorities have had to work together and overcome local jealousies and rivalries. The result of this could mean that by 1980 the Tyne might once more become a river suitable for fishing.

The case of the Tyne is by no means exceptional, nor is it the worst. Pollution of the Trent is now so serious that in May 1970 the Medical Officer of Health for Burton-on-Trent advised the organisers of the annual raft race to cancel the event in the interests of the safety of the competitors. In the whole of Europe only the Rhine is considered more polluted than the Trent. Major contributions to the pollution of the Trent come from its numerous tributaries including the Tame and the Idle. The sources of the contaminants are untreated sewage, detergents, pottery wastes, poisonous chemicals and domestic rubbish. In fact everything undesirable happens to the Trent.

It is impossible to exaggerate the value of a river supporting fish life as this is a very good guide to the purity of the water. As a river becomes more and more polluted there is a corresponding decrease in the number and species of fish that can survive in it. Equally when the pollution in the river decreases one of the first indications of this is a return of fish life to the water. On this basis it is possible to understand the enthusiasm that has greeted the news that in both September 1969 and July 1970 bleak were caught in the Thames although none had been caught prior to those dates.

Fish are driven away or killed by pollution in several ways: they may be poisoned by metals and metal salts such as copper and lead; they may be poisoned by organic substances such as insecticides, or they may become unable to obtain enough oxygen from the water because of a lowering of the oxygen level by other substances or by an increase in the temperature of the water. It is not enough just to ensure that no poisons enter the water, as many waste materials such as sewage and food residues are not poisonous to fish but do use up the oxygen during their decomposition. In this way the water becomes deoxygenated and less capable of supporting life.

Effluents need to be checked not only for acidity, dissolved solids and suspended matter but also for their B.O.D. (Biochemical Oxygen Demand). The B.O.D. is the amount of oxygen, in milligrammes, that a 1,000-millilitre sample of the effluent will take up when kept for five days at 20° C. Because of the delay in producing the B.O.D. figures, which could prevent any necessary emergency treatment for a river from being rapidly applied, a quicker method for estimating the oxygen requirement has been developed. This is based upon the uptake of oxygen by the sample during a four-hour period from a chemical oxidising agent called potassium permanganate.

Obviously the greater the amount of oxygen the waste takes up the less will be left to support aquatic life. Fortunately for the fish and the purity of our rivers, oxygen is being continuously dissolved from the atmosphere into the water as it passes along. The greater the degree of agitation, as in passing over a waterfall or weir, the quicker the uptake of the oxygen. This can always be helped by artificially aerating the water or effluent by pumping air through it or by making it pass over man-made waterfalls. There is a limit to the amount of oxygen that can be dissolved at any particular temperature and it is unusual for the oxygen level actually to attain the saturation level. An analysis of the amount of oxygen present at a given temperature can be expressed as a percentage of the possible saturation figure. A river showing a continuously low dissolved oxygen concentration is usually heavily polluted. This means it is not only incapable of supporting fish but also incapable of breaking down and oxidising the impurities that have been poured into it.

Another effect that lowers the amount of oxygen is heating the water. The amount of dissolved oxygen decreases as the temperature of the water increases, and just a small increase in temperature of about 10° C leads to a reduction of 20 per cent in the maximum level of dissolved oxygen. The discharge of warm effluents into rivers is referred to as thermal pollution of water and is a relatively common feature of many industrial wastes. A wide variety of fish can only exist when the dissolved oxygen level is about 4–5 parts per million and the water is neither very acidic nor very alkaline. It is failure to meet these fairly moderate requirements that has led to high levels of pollution and the elimination of aquatic life in many rivers.

Fish vary greatly in their susceptibility to thermal pollution. For trout the lethal temperature is about 25° C, whilst goldfish and perch can tolerate temperatures above 30° C. Most river authorities in the United Kingdom have specified upper limits for effluent temperatures, and the Federal Water Pollution Control Administration has made similar recommendations in the U.S.A. By far the worst culprits in this are electricity power stations which, unless carefully controlled, can easily raise the temperature of the river by several degrees. This problem of thermal pollution is particularly acute during heat-waves, when the natural temperature of rivers is likely to be high. Fortunately modern power stations are usually constructed with adequate cooling towers and can normally return the water to the river below 30° C, but ideally much lower temperatures are required if fish-spawning is to take place. The problem becomes even more acute when the increased temperature is combined with poisons in the river as the sensitivity of the fish to the pollutants is increased under these conditions. It is the cumulative effect of the different forms of water pollution that must be combatted before substantial improvements are likely to be seen in the most grossly contaminated rivers.

One of the greatest sources of pollution of all is the discharge of both untreated and only partially treated sewage from many towns and cities in practically all countries in the world. This is such a large subject that it has been dealt with separately in Chapter 5. The other major source of pollutants is industry, with a different problem being created by each individual manufacturing process which involves the use of water.

If pollution of rivers and estuaries is to be prevented it is obvious that methods for the disposal and treatment of water-borne trade wastes must be encouraged and developed on a much greater scale than has hitherto been done. In many cases it is possible for industrial wastes to be discharged into the public sewers for treatment at the local sewage works. Under the Public Health (Drainage of Trade Premises) Act, 1937, manufacturers were given the legal right to discharge effluent to the sewers as long as it met the standards laid down by the local authorities. This requires special agreement with the local authority, and often necessitates pre-treatment at the factory to remove any particularly unpleasant materials that are beyond the scope of the sewage works. Failure to carry out necessary pre-treatment can lead to difficulties of the type experienced by the Greater London Council. In one case they had a sewer blocked by discharges from a ready-mix concrete plant, and in another case found that a blockage was due to a food factory having discharged five tons of edible (but not soluble) fat down the drains. There are now well-established procedures for dealing with nearly all industrial wastes and little excuse exists for companies which are still discharging undiluted and untreated material into the rivers.

Because of the increasing costs of water, many companies have developed methods for re-cycling and re-using much of the water employed in their processes. This is particularly so in coal-washing plants; the dirty washing water is allowed to settle out in specially designed tanks called lagoons. Where necessary, coagulating chemicals are added to assist the settling of the particles before the water is re-circulated to be used again for washing. The settled waste, consisting mainly of coal particles, is dried and used as a fuel. Similar recirculation processes are used in canning factories and pulp and paper mills.

In certain cases treatment of industrial waste liquors can lead to recovery of useful and often expensive raw materials. Even relatively simple techniques like screening of wastes in paper mills can lead to the recovery of substantial amounts of paper fibre. Reclaiming of materials is of particular value with electrolytic liquors which contain quantities of acid in addition to useful metals such as chromium, copper and iron. The chromium plating of steel actually involves three separate plating processes requiring twenty-one operations of which eleven involve rinsing with large quantities of water. Simple economic considerations justify the re-cycling of as much of this water as possible. For this purpose, extensive use is made of ion-exchange resins of the type employed in water-softeners. These remove the contaminants almost completely from the rinsing water by exchanging them for other substances on the resin. The valuable metals removed by the ion-exchange process can later be isolated from the resins. In most cases continuous re-cycling is possible with only periodic topping up to replace evaporation. Purification and re-cycling processes are similarly used in such widely differing procedures as canning and car-body sanding. Heavy demands on local water supplies can thus be avoided by industrial re-use methods, and contamination of rivers and sewage correspondingly reduced.

It is still common for some trade wastes to be used for irrigation purposes. The distribution of waste in this way is intended to lead to breakdown to innocuous substances owing to the action of bacteria in the earth, in addition to supplying nutrient to the soil. This method of disposal has been used for dealing with waste from the preparation and packaging of dairy products, and has been applied substantially in New Zealand and the U.S.A. The type of waste that can be treated in this manner must obviously be of some nutritional value to the soil and great care is needed to ensure that toxic chemicals and substances that will kill off the soil bacteria are absent.

Similar attempts have been made to use the ammoniacal liquors from coal-gas manufacturing plants as a substitute for solid nitrogenous fertilisers. Whilst a certain amount of success has been achieved with this, it is necessary to ensure that the liquor is free from coal tars and toxic thiocyantes to avoid destroying the very crops it is intended to help. There is no doubt that the intelligent use of liquor wastes in this way can not only solve some of the disposal problems but also act as a useful method of both irrigation and fertilisation.

Although the discharge of waste into disused mines and pits has been a fairly common method of disposal it suffers from several serious disadvantages, and there are now a number of laws restricting disposal in this manner. Under the Water Act, 1945, it is an offence to dispose of effluents into the ground if there is a possibility that water supplies will become polluted. Further controls on this were applied by the Water Resources Act, 1963. These restrictions were made because cases arose in which wastes which had been apparently safely disposed of down mineshafts finished up by passing into underground streams and eventually entering domestic water supplies. There have been instances in which obnoxious materials have travelled great distances underground before reappearing several years after disposal. In one case wastes from an explosives factory which had been dumped into chalk borings during World War I were found to impart a yellow colour to well water many years later.

As restrictions on dumping and direct discharge into rivers increase, a progressively greater amount of trade waste will pass to sewage works for treatment. Whatever happens, there will be a greater obligation on industry to pre-treat its own waste in the factory site. This is obviously the best place, as the manufacturers will know, or can easily establish, the nature of the effluent and the most toxic constituents. In some cases dilution of heavily polluted liquid waste with less contaminated water will be enough for the effluent to be acceptable for the local sewers. Where this is not possible special plant is often necessary for the various purification steps to be carried out.

'On site' purification, prior to sewage or river discharge, must set out to achieve certain objectives, including the removal of toxic materials, neutralisation of acids and alkalis, cooling and re-oxygenation of the effluent and removal of suspended solids. These requirements can be reached by treating the effluent in several stages. Suspended solids can be removed by slow passage through settling tanks before being treated with chemicals to precipitate or destroy toxic materials.

The next step is neutralisation, probably followed by treatment with coagulants and filtration. If the effluent has become heated above normal temperatures during these stages it will need to be passed down a cooling tower, or through some form of partial refrigeration, and aerated to increase the amount of dissolved oxygen before it can be discharged.

Treatment of all industrial wastes in this way would soon improve the quality of many of the world's really heavily polluted rivers. Purification of discharges to rivers is in the interests of industry itself. Currently many factories are having to treat the water specially when they extract it from the river before use because it has been polluted further upstream by someone else. Whilst this might seem to be poetic justice there should be some scope in this type of situation for co-operation and shared costs to prevent the pollution in the first place.

Farming creates its own particular problems of pollution from the run-off of animal wastes, fertilisers and pesticides into water courses. It has been calculated that cows produce ten tons of solid waste per head in one year; this is about four times the amount from an adult human being. Fortunately much of this waste is usefully applied by being evenly distributed over the land for cultivation. Farm wastes can be discharged into the sewers, like other trade wastes, but if undiluted can result in disruption of the normal regular working of the sewage treatment plant. Various suggestions have been made that further study is needed of the methods for dealing with farm wastes, particularly since the advent of factory farming methods, which are not always accompanied by crop growing and the need for manure. One proposal has been that farms should install small plants for the biological treatment of the waste in lagoons. This method, however, can give rise to unpleasant odours and adds to the water pollution. Composting is an alternative disposal method used for initial treatment before the waste is distributed over the land. In cases such as this, in which large quantities of animal wastes are piled together, special efforts must be made to ensure that drainage from the compost to the river does not occur.

Agricultural fertilisers have given rise to a certain amount of concern because of the accumulation of inorganic nitrates and related chemicals in groundwater. As was mentioned in the previous chapter, nitrates in drinking water are metabolised by the human body to nitrites, and can give rise to an illness known as methemoglobinemia in infants. This problem is of special concern in countries in which there is a natural high nitrate level in the water and can be introduced in areas in which there is a rapid rainwater drainage from heavily fertilised fields. The recommended limit for nitrates in drinking water, established by the World Health Organisation, is 50 milligrammes per litre, although a level between 50 and 100 milligrammes is considered acceptable. But some countries feel that this is too high, and have fixed much lower levels. Currently, Britain follows the W.H.O. recommendations, and if nitrate levels in drinking water show any indication of reaching undesired levels, alternative sources of domestic supplies are used.

Methemoglobinemia may occur in people of all age groups, although most cases arise with children under one year old. It results from the failure of blood to carry sufficient oxygen around the body owing to conversion of active hemoglobin into inactive methemoglobin. Since the illness was recognised clinically in 1945 about 2,000 cases have been reported, and most of the fatalities have been young children. Countries as wide apart as U.S.A., Czechoslovakia and Russia have all reported instances of children having above-normal methemoglobin concentrations in their blood.

Particular concern exists in Israel, because many wells providing drinking water have shown a constant increase in nitrate content over several years. In some areas the 50 parts per million level is already exceeded. As the illness and its effects are not yet fully understood measures to combat it are still being developed. Analysis of water for both nitrates and nitrites is the first step in controlling it, and chemical studies are standard practice. Where levels close to 50 p.p.m. exist it is necessary to reduce the quantities of nitrates ingested from other sources. Such foods as rhubarb and spinach, with natural high nitrate levels, are cut out of the diet. Fortunately the illness can be combatted by regular doses of Vitamin C, and it may well be that it is still relatively uncommon because it is a normal procedure to give foods containing this vitamin to children. Should nitrate levels continue to increase it will become essential to treat drinking water by special methods to reduce these to more acceptable levels.

Of even greater concern is the widespread and often indiscriminate use of pesticides which so frequently finish up in water and get transferred from one form of life to another. The pesticide problem, which has been treated in considerable detail by a number of authors, really started with the widespread use of D.D.T. This was first introduced on an experimental basis in 1939 and came into commercial use in 1946 after being successfully used during World War II. Since those days the number of pesticides has greatly increased and now runs into many hundreds. Despite the apparent benefits that they have brought to man they have carried in their wake great dangers. The U.S.A. Department of the Interior started surveys on the death of fish caused by pollution as far back as June 1960, and has produced an annual report since that time showing how and where fish have been killed in the U.S.A. Once a pesticide enters the water it may not only lead to serious changes in the ecology of the river but may eventually enter the bodies of human beings. It can do this in two ways, either by passing through the water purification process in an unchanged form, or by being stored in the tissue of aquatic life later eaten by man. The problem has arisen because certain classes of the pesticides that have been manufactured do not break down very quickly when they have been sprayed over the land and on plants. These are known as persistent pesticides, and in many instances continual spraying over several years has led to a substantial build-up of the pesticide residues on the land and in the rivers.

Despite the restrictions on the use of certain pesticides it is still common to read of serious pollution problems being caused either by carelessness or because of a deliberate disregard of the instructions. In June 1969 thousands of fish were killed in the River Rhine and the cause was attributed by some scientists to an insecticide called endosulphan. The Rhine is in any case a very heavily polluted river, and this particular tragedy served to draw attention to the condition of the water which on this occasion necessitated restocking the river with large numbers of young fish. One long-term benefit is that it has resulted in much more concern over the state of the Rhine, and tighter controls are being implemented that will probably help to produce a greatly improved river within ten years. A similar situation occurred in 1970 in Chichester harbour, when part of an aerial crop-spray of phorate granules was carried over the sea by a breeze. Within a day of the spraying the foreshore was covered with dead mullet. In the U.S.A. a farmer treated a pile of corn with the chlorinated hydrocarbon insecticide dieldrin. The corn was stacked along the edge of a river and a heavy rainfall washed a lot of the active chemical into the water. This resulted in the deaths of over two thousand fish.

The number of such cases in many countries over the years runs into thousands. The first steps to preventing frequent massacres of fish populations have been taken. Many of the more dangerous persistent pesticides have been banned in countries such as Sweden and Hungary, or their use has been restricted to short periods during the year. There is not at present much uniformity on this. In the United Kingdom restriction was originally limited to a voluntary control by manufacturers and users until 1969, when some controls were made compulsory. Even so, the attitude in this country is considered by various action groups to be very liberal and efforts are being made to obtain a total ban on organochloride pesticides. Special regulations have also been imposed to ensure that cans of dangerous chemicals are better labelled to encourage and assist in correct handling. More particularly the chemical industry has taken a long, hard look at the nature of the compounds it has produced and is making pesticides which are less persistent and more selective with their toxicity. Action of this type may serve in the long run to retain many of the advantages of pesticides, which have after all contributed to an increase in the world's food supplies, and yet rid us of their disadvantages.

Pollution in rivers is something that most industrialised nations have begun to take seriously. Action that has already been taken is having its effect and acting as a spur to areas in which nothing has been done. It is no longer fair to refer to the Thames as an open sewer, although the description is still apt for many other major rivers. What has happened to the Thames is certainly one of the great stories of pollution control, as it had showed a progressive deterioration in quality from the beginning of the century to the late 1950s. Other rivers too have shown that if positive steps are taken they can be improved. Both the Dee and the Colne have been reported to be cleaner as a direct result of restrictions on the discharge of pollutants.

It is, however, worth looking at the story of the Thames in more detail. A hundred years ago Members of Parliament found the stench from the Thames so objectionable that both Houses were adjourned. It has been estimated that from about 1920 onwards, until recently, virtually no aquatic life existed from Fulham to the estuary, and in 1949 it was established that during the summer more than twenty miles of the tidal part of the river contained no dissolved oxygen. The recent improvement that has taken place has come about by the expenditure of £20 million since 1949 by the former London County Council and the now Greater London Council, coupled with greater power as a result of the anti-pollution laws passed by Parliament during the previous ten years. A great deal of this money has gone on improved sewage works, as most of the treated, and much of the untreated, sewage from the London area is discharged into the river. By 1973 improvements at two large sewage works alone will have cost £26 million. Restrictions on the discharge of wastes from ships in the docks and close co-operation with the Port of London Authority inspectors have all contributed to the improvements. Pollution from industry is being steadily reduced and regular checks are made to prevent any surreptitious discharging of pollutants. The various authorities in the Thames valley are justifiably proud of the changes that have been made within a period of ten years, but nobody is complacent. Many improvements still need to be carried out, particularly to the tributaries of the Thames. As recently as February 1970 both the Wandle, which flows into the Thames at Wandsworth, and the Lee (of Isaac Walton fame), which meets the Thames at Blackwall, were the subjects of scathing articles by Arthur Hawkey in the *Evening Standard*.

Much of the investigation on the state and nature of British rivers is carried out by the Water Pollution Research Laboratory. It is not commonly known that this laboratory, now part of the new Department of the Environment, was originally established as long ago as 1927 with the function of investigating pollution of natural waters, sewage and trade effluents. Part of the cost of the work carried out is directly offset by contributions from industry. The extensive work of the laboratory is described each year in an annual report and many techniques devised and pioneered there have become widely used. It is unfortunate that the contribution made by the Water Pollution Research Laboratory over the years to keep rivers and estuaries clean is so little known and appreciated. Much of the data collected and methods investigated are now the basis for many improvements scheduled for water pollution control not only in the United Kingdom but in many other parts of the world as well.

Many of the delays which have occurred in dealing with pollution have arisen from the inadequate laws which have received little attention until quite recently, and it is only during the past few years that water authorities have had adequate powers to deal with offenders.

This is not to say that no earlier attempts were made to pass laws on the matter. The Public Health Act, 1875, which arose as a result of the cholera epidemics in the mid-nineteenth century, specified certain standards for treating sewage before it could be discharged into the rivers and was followed a year later by the Rivers Pollution Prevention Act, 1876. When originally passed, these Acts were revolutionary. Unfortunately the technical knowledge necessary for them to be effective did not exist at that time. In addition, when they were put into practice there were many loopholes which enabled polluters to evade the intended controls. The 1876 Act, however, was the main law against water pollution until the middle of this century. Concern over water pollution was beginning to grow in the 1930's and in 1949 the Ministry of Health published a sub-committee report entitled 'Prevention of River Pollution', which contained a number of recommendations, many of which were incorporated into the Rivers (Prevention of Pollution) Act, 1951. This Act restricted the discharge of any new wastes into the rivers and was later extended in 1961 to cover wastes that had been freely discharged prior to 1951. These controls were further strengthened in 1963 when the administration of inland water courses was completely re-organised under the Water Resources Act, which established a Water Resources Board covering twenty-nine Regional River Authorities in England and Wales and ten River Purification Boards covering Scotland.

As a direct result of these changes the steps necessary to stop pollution and clean up the rivers can be carried out much more easily. This in itself is an encouragement to local councils to take steps to avoid polluting the rivers with sewage. Public opinion has now come out strongly in favour of cleaning up the rivers and local residents are thus less likely to object when rates are increased to build a new sewage works or to improve an old one. This is reflected in the figures for annual expenditure on new sewers and sewage works: in 1948–9 it was £9 million; in 1958–9, £30 million; and in 1968–9, £110 million. Great Britain was the first country to suffer from the detrimental by-products of the industrial revolution; it is now one of the pioneers in the field of water pollution prevention. This position and continual improvement in the quality of inland waters will only be maintained if we are willing to spend even more money to remove the poisons which could eventually make all water undrinkable and undo the results that have already been achieved. We cannot transfer the blame for our river pollution problems on to other nations, as is possible on the European mainland. The difficulties are of our own making and we must find our own solutions.

4 The battle for the beaches and oceans

For many years the oceans have been treated by all nations as if they are vast bottomless pits that can take an unlimited amount of waste discarded by so-called civilised man. Recently, however, it has become quite obvious that some of the substances discharged into the oceans do not break down rapidly and are often regurgitated upon the beaches by the action of winds and tides. Visible signs of this have been seen in the increase of oil slicks on the water and oil residues on the beaches, coupled with a substantial rise in the number of dead sea-birds washed on to the shores.

The rapid deterioration that has occurred in the situation has been shown in the surveys carried out by the Coastal Anti-Pollution League. Since 1960 the League has published two 'Golden Lists' of clean beaches in the United Kingdom, grading those included with from one to four stars. These lists have been criticised for drawing attention to the many dirty beaches which are excluded. In May 1970 the list publishers claimed that some local authorities were honest enough to admit that their beaches were filthy. In addition to this the International Beached Birds Survey, carried out in Great Britain by the Royal Society for the Protection of Birds in co-operation with a number of other organisations, found over 3,000 dead birds in 950 miles of coastline in just one week-end. In the first three months of 1970 over 22,000 dead birds were washed up on the coast. This was the highest recorded figure for any three-monthly period and cannot be attributed to bad weather conditions. The situation is now so bad that the R.S.P.B. has adopted a policy of only attempting to clean up lightly oiled birds. Little or nothing can be done for birds poisoned by chemicals and the more seriously oiled birds have to be destroyed humanely.

Even more moderate releases of oil are responsible for substantial numbers of bird deaths. In February 1969 150 tons of oil released off the Dutch Frisian Islands caused the deaths of 20,000 birds. Similarly the holing of the *Hamilton Trader* led to the escape of 700 tons of oil, which affected beaches along the coast of North Wales and North-west England and caused the deaths of 1,500 sea-birds.

With over 4,000 tankers in use throughout the world carrying 600 million tons of crude oil, accidents are only too common. Most maritime nations have now had to face the experience of dealing with polluted beaches from oil slicks. People in Puerto Rico, the Bahamas, Germany, Holland and France all know what it is to have their livelihoods threatened because of some accident or downright carelessness that has let to black, tarry, crude oil spreading along their coastlines. Shellfish beds have been ruined, sea-birds have been decimated and holiday resorts heavily polluted.

Unfortunately relatively little work has yet been done on the dispersal and effects of oil slicks. It is only in recent years that investigations have really got under way, and it may be some time before a lot of the results are available. One point that has emerged from studies by the former Ministry of Technology is that the amount of oil that can be safely discharged when spread over one mile, with dispersion occurring within three hours, is only thirteen gallons. This amount is minute when compared with the vast quantities carried in the supertankers and is certainly no encouragement for ships to dump oil waste at sea.

The main problem with oil is that it floats on the surface of the water and can be carried great distances until it either adheres to something solid, like the beach or rocks, is chemically broken down by the combined action of water, oxygen, sunlight and substances dissolved in the water or until it sinks due to evaporation of volatile components. These processes can take some considerable time, and thus oil slicks persist at sea for weeks and even months, whilst the tarry deposits can last almost indefinitely on rocks and breakwaters.

Most of this oil pollution stems from the discharge of tank-cleaning residues after the tanker has discharged its cargo at the refinery and is returning to the oil field. The amount of oil that used to be dumped as washings in this way was enormous despite the individual controls that many nations applied within their own territorial waters.

Discharges at sea create an international problem and, unlike inland waterways, no country can solve its polluted beach and dead bird problems unilaterally. Beaches around Great Britain have suffered from oil pollution for years despite the fact that it is illegal for a ship to discharge oil within the three-mile limit and despite an international ban on dumping in the English Channel and the North Sea.

In fact very little progress was made until the 'load on top' technique was devised and internationally adopted. This procedure, which is now used extensively, is to pump all the oily waste sea-water washings into one tank in the ship and to allow the oil to settle on top. The lower layer of water is pumped into the sea and the oil residue retained in the tank until a fresh load is taken on board. This technique has recently been improved by the more progressive and socially minded oil companies. They have had a special small oil tank fitted into the hulls of their large tankers just to accommodate up to 40 tons of oil-waste left after washing down. This waste can be separately treated or unloaded when the ship has docked again.

International agreements on shipping are made by the International Governmental Maritime Consultative Organisation and the agreement dealing with oil, the International Convention for the Prevention of Pollution of the Sea by Oil, can only be amended by a two-thirds majority decision. Even when that has been done, each individual member country still has to give parliamentary approval to any agreed change. As a result it can take three or four years for any substantial alteration to be effectively implemented on an international scale.

Currently there is an attempt to obtain international agreement to impose additional restrictions on oil discharges at sea. These include extending the minimum distance from the shore to fifty miles and limiting any total discharge to 1/315,000 of the cargo carried. Already the discharging of oil is banned in the Mediterranean, the Baltic and the North Sea as well as in much of the Atlantic Ocean and off the coast of North America. The major oil companies whose tankers carry 80 per cent of the total world tonnage are only too pleased to help reduce pollution and have often instituted research to improve control methods. Most responsible companies will operate to the new standards even before full international agreement is reached. The real problem is with the shipping that runs under the so-called 'flags of convenience' – an appropriate name for ships which have the reputation of using the sea lanes as a ship's convenience for any waste they care to discharge. For irresponsible people like this, the only answer is increased vigilance and heavier penalties for those who are caught.

As the *Torrey Canyon* incident represented one of the worst cases of oil pollution occurring in any country in the world, it is worth looking at in some detail. It also gives some idea of the legal complexities which can arise from international shipping. The S.S. *Torrey Canyon* carried 119,000 tons of Kuwait crude oil and at the time of the accident on 18 March 1967 was one of the largest ships afloat, being only 57 feet shorter than the *Queen Elizabeth*. It was carrying oil for British Petroleum (a company in which the United Kingdom Government is a substantial shareholder) under a Liberian flag and was owned by a company registered in Bermuda. Despite the fact that the accident occurred between the British-owned Scilly Isles and the British mainland it was in fact in international waters. These complications alone indicated possibilities of many years of legal wrangles, particularly after the *Torrey Canyon* was finished off by bombing and its oil had ruined miles of beaches in England and France. Eventually the matter of compensation was settled out of court in 1969. Each of the two nations involved received one and a half million pounds and an additional twenty-five thousand pounds was paid for damage to private property. The compensation only represented about half the cost of fighting the pollution and the relatively quick settlement was expedited by the British Government nailing a writ on the *Lake Palourde*, sister ship to the *Torrey Canyon*, in Singapore. The French Government emulated the feat when the ship later docked at Rotterdam.

The *Torrey Canyon* disaster provided an opportunity to try out even the most hairbrained ideas for removing the pollution and indicated the total inadequacy of the research which had been carried out up to that time for dealing with major oil discharges. As a result, money was spent on applying untried methods, many of which were unsuccessful. The major expenditure came in using detergents for cleaning beaches and quays. Whilst apparently helping to improve the appearance of the polluted areas the detergents wreaked havoc amongst the flora and fauna of the ocean. It has been claimed, from results in France, that there was a quicker recovery from pollution in untreated areas than in those cleaned with detergents.

Despite the world-wide publicity given to the effects of oil dumping the practice still goes on, and oil pollution around the coast of Great Britain is becoming worse every year. It has been estimated that over 100,000 tons of oil are spilt into the waters around the British Isles every year despite the restrictions already existing. Fortunately all aircraft are now instructed to keep a close lookout for ships trailing an attendant oil slick behind them. The result of this co-operation has been a steady increase in successful prosecutions from fifty-nine in 1967, to sixty-two in 1968 and seventy-seven in 1969. Total fines imposed in 1969 amounted to £17,550, averaging only £225. These seem completely inadequate when it is realised that it cost Southend Corporation alone £1,500 to clear the sea front after a tanker cleaned out its tanks in the Thames Estuary.

For all this there are still many difficulties in detecting the polluters, particularly those who choose to do so under cover of night. It has been suggested that the best way to cope with this would be to 'label' each cargo of oil with some form of radioactive tracer or characteristic dye. So far no practicable method for doing this has been adopted. Although the basic idea would certainly answer the problem there are a number of obvious difficulties and disadvantages. Present procedures are to take a sample of the oil in the slick and subject it to physical and chemical analyses in an effort to determine its origin.

As prevention is not always possible, and even with 100 per cent international control and agreement, accidents will inevitably still occur, a great deal of the hope for clean beaches still lies in being able to deal with the leakages and discharges after they have taken place. Probably the greatest outcome of the *Torrey Canyon* disaster was that it showed that nobody knew how to deal with a major oil release. It certainly served as a spur to research in this field and has led to a number of new approaches. As a result the British Government was much better prepared to deal with any pollution arising from the collision between the two Liberian-registered tankers *Pacific Glory* and *Allegro* in October 1970. Fortunately only a small proportion of the oil was actually released and a major disaster was averted.

In many cases the direct method of treating oil slicks with detergents has been employed. This is usually effective, but the large quantities necessary are not only expensive but probably cause as much damage to marine life as the oil would if left alone. A Warren Spring Laboratory Report in 1963 pointed out that all effective detergents at that time available for dealing with oil residues are toxic to marine life, particularly to shellfish. As recently as April 1970 local coastal authorities in the southern county of Dorset were warned not to use detergents for removing oil from beaches because of the increased danger to wildlife. Some areas were advised to rely on the natural cleaning action of the sea.

Attempts have been made to skim the oil off the surface of the water and this has been successful on a limited scale. The most encouraging treatment for large oil slicks has been a development by the Shell Petroleum Company, demonstrated for the first time in April 1970. The method has been called the 'Sand Sink System' and involves spraying the slick with a slurry of specially treated sand. On sinking, the sand carries the oil to the sea bed. The procedure was claimed to be 95 per cent successful in clearing a slick of 100 tons of crude oil in 45 minutes. The spraying is carried out from forty nozzles fitted to two special booms each 60 feet long directed at right-angles to the hull of the spraying vessel. Some critics of spraying and sinking methods claim that results are only temporary and that the oil is slowly released to the surface over a period of time. This need not be the case if the oil becomes chemically combined or physically absorbed by the sinking agent.

Plastic booms used for containing oil slicks within a limited area have been found to be of limited use mainly in calm waters as any substantial rise and fall of waves enables the oil to escape. When the oil from the *Torrey Canyon* hit the French coast use was made of methods which were not applied in Britain. Twenty thousand tons of oil were sunk at sea by spraying with 3,000 tons of chalk. Oil which reached the beaches was absorbed by straw and sawdust at the high tide limit and then ignited before it could spread everywhere. Their third approach was to collect the oily water mess and dump it into pits dug in the earth. By this means the two components slowly separated. After the water had drained away the oil was burnt. It was estimated that 6,000 tons of oil were recovered and used in this way.

When a major accident occurs, as when the *Monte Ulia* rammed the refinery jetty at Coryton, speed is essential to deal with the spillage before it can spread too far and actually reach a beach. At present the Department of Trade and Industry is only responsible for dealing with oil slicks more than a mile off-shore and local Principal Marine Officers have powers to co-ordinate the efforts of the armed forces and coastguards to deal with the emergency situation. Oil slicks nearer in-shore must be cleared by the local authority. This division of responsibility is obviously undesirable particularly as the local authority is only obliged to deal with those slicks likely to foul publicly used beaches. An arrangement of this nature can only lead to delays in taking action in the hope that the threat will go away to another local authority's area.

It would appear that what is needed in the United Kingdom is a national body with both the facilities and the power to take immediate action necessary to handle any large oil slick approaching the coast before it hits the shore anywhere and then to continue the action if it does manage to reach the beaches. There is also a need for an extension of British territorial waters to achieve continuous control over sea-lanes operating between the mainland and off-shore islands. Had the British territorial limits extended twelve miles from the land the *Torrey Canyon* would have been inside the limit and subject to British maritime authority, as would have been the *Pacific Glory* and the *Allegro*.

Whilst increased international co-operation will help to reduce the deliberate discharge of oily wastes, the only real solution lies in the provision of adequate residue treatment facilities at every unloading port. Currently these are few and far between. Nations that are truly concerned about the environment should make the installation of oil-waste storage plant obligatory at all depots. In addition, tankers not using the 'load on top' system should be blacklisted by the major oil-consuming countries and not allowed to discharge cargoes until able to show that they use this method.

It would be a mistake to assume that all the fouling of the beaches occurs from oil. A great deal of it is man-made from sewage outlets, and occasionally from dumped chemicals. Eighty per cent of the coastal towns in Great Britain discharge their sewage into the sea and this constitutes 25 per cent of the national total. A large proportion of this is untreated sewage and very little of the remainder can be considered to have had adequate treatment. Some of the outfalls from which the discharge is made are only a few hundred yards from beaches used for bathing. These can usually be identified by the flocks of gulls hovering just beyond picking up scraps.

This easy method of disposing of human wastes is a common practice in most countries. Under carefully determined conditions coupled with initial screening and settling treatment it can be acceptable, but all too frequently this is not the case. Large-scale studies on the water in Raritan Bay, between New Jersey and New York, showed that the daily discharge of 480 million gallons per day of industrial wastes and domestic sewage have made the water unsuitable even for industrial use and have tainted fish and shellfish. In addition more than a quarter of the samples of clams from the bay showed the presence of the dangerous *Salmonella* bacteria.

Advocates of this method of disposal usually claim that there is a negligible health risk to people who swim in the sea in these areas. To most people the very concept of swimming in sewage-polluted water is repugnant, although a very large number in their ignorance have done so. The obvious disadvantages of this procedure are that the sewage may be carried back to the shore and may easily lead to swimmers encountering unpleasant waterborne solids. Objections of this type have not prevented some local authorities from continuing and extending the use of this method of disposal.

By carefully plotting the tidal flow and wind effects marine disposal can be achieved in a reasonably acceptable manner if the waste pipes extend far enough out to sea. This usually means from one to four miles, as most complaints arise in resorts where sewage outlets are only three or four hundred yards from the shore. In Tel Aviv, Israel, acceptable dilution and dispersion in the sea has been achieved by a sewage pipe with its outlet about half a mile from the shore.

In Great Britain the Institute of Sewage Purification published a list of recommendations in 1963 to establish criteria for discharging sewage into tidal waters. These included resiting or extending the outfalls, preliminary treatment and even full treatment in some cases to bring about improvements. At the same time the Institute pointed out that 'it should be remembered that inland towns are required to treat sewage and trade effluents to a high degree of purity in order to protect inland waters. It should be the pride of a seaside resort attracting holidaymakers that its sewage is adequately treated before discharge to tidal waters.'

Fortunately there are some towns that are sufficiently concerned about the quality of their beaches and the reputation of their standards to be prepared to instal full treatment plants. Bournemouth is a case in point. Tourism to this south coast town is worth over £20 million per annum, and in an effort to maintain its reputation for clean, golden beaches it is in the process of spending £5 million to construct an inland sewage works to take all the town's sewage, instead of discharging most of it into the sea, as is currently done. The greatest lever on coastal resorts to do this type of improvement is public opinion. Many residents of these towns would be horrified to learn of the archaic method by which their sewage is treated: unfortunately too few know. Any legislation to deal with the situation still appears to be a long way off, so we will still be swimming in sewage for years to come.

Unfortunately, whilst holidaymakers may be the source of prosperity for coastal resorts they have no control over the towns' purse strings. Only by refusing to frequent contaminated areas will the necessary measures be taken to attract back the visitors.

5 Dealing with sewage

It is unfortunate that the very name sewage is enough to cause people to turn away in contempt or embarrassment. This approach to the subject has meant that it is usually inadequately discussed in public conversation and even in local council chambers. As populations increase and cities grow larger it becomes even more important that human wastes are fully dealt with to prevent unpleasant odours, avoid water pollution and prevent the spread of diseases. The treatment of sewage should be of major concern to everyone.

When populations were both small and mobile it was quite sufficient for man to relieve himself on the ground and walk away, leaving the processes of nature or scavenging animals to break down his wastes. As communities became larger this had to be carried out in a rather more organised manner as, we are told in the Bible, was required of the Israelites during the years spent in the desert. They were instructed by Moses to carry out their bodily functions outside the limits of the camp and were to dig a hole for this purpose and cover it over after use. This was adequate for a group of people constantly on the move and left the breaking down of the organic material to bacteria in the soil.

Once man settled into permanent dwellings it was necessary to develop more sophisticated methods of disposal. As a result some ancient civilisations constructed sanitary systems superior to those in operation in some of our modern towns. Water-flushed latrines existed in the Babylonian empire as far back as 2,500 B.C., and the Aegean civilisation which existed between 3,400–1,200 B.C. developed an elaborate system of drainage which can still be seen at the Palace of Minos on the island of Crete.

Development of decent sewage disposal systems in Great Britain did not start until well after the industrial revolution. Prior to the nineteenth century, cities in Europe were foul, unhygienic places in which refuse, trade wastes and sewage were often tipped into the streets, causing appalling smells and helping the spread of diseases. More civilised methods of dealing with wastes at an earlier stage would undoubtedly have saved Europe from many of the plagues which were able to spread rapidly under such disgusting conditions. Even when the waste and excrement were cleared from the roads they were often thrown into the nearest river in the hope that they would be carried downstream and out to sea.

This disastrous situation was further aggravated when factory wastes were added to those already going into the rivers. As cities increased in size more and more human wastes were discharged in this way with little or no treatment. Owing to the deterioration of the rivers and considerable public pressure the British government of that time set up two Royal Commissions, in 1865 and 1868, to look into the extent and nature of waste disposal in general. But for sewage itself there was a further wait of thirty years until the Royal Commission on Sewage Disposal was appointed in 1898. The extensive study carried out by this body filled nine volumes and the work lasted well into the twentieth century. It served as the basis for a significant advance in dealing with large quantities of sewage and established many standards which still apply today.

Many of the traditional methods of disposal for sewage are still used in under-developed countries and even in less enlightened sections of some of the wealthy nations. Disposal of crude sewage into rivers does not need to be either undesirable or unpleasant if certain precautions are carefully observed. All rivers have a natural capacity for self-purification due to the presence of dissolved oxygen and the action of aerobic bacteria. If limited quantities of sewage are first screened to remove large suspended solids and then added to the river at a rate that leads to a five hundred times dilution no unpleasant effects should occur. There are, however, insufficient rivers in most industrialised countries for this to be a viable method of disposal and the effects of overloading the rivers are only too obvious throughout the world.

Once the oxygen in the river has been exhausted the biochemical reactions taking place change as anaerobic bacteria (bacteria that do not use oxygen) then start to decompose the sewage and give rise to the awful smells which are so common with heavily polluted rivers. For this reason disposal of sewage in this way is undesirable for all but the smallest community sited by the side of a large, fast-flowing river.

An alternative method of dealing with sewage that has been used extensively throughout the world is the use of sewage farms. Basically the method consists of irrigating the soil with screened sewage from which the coarse solid material such as paper and rags has been removed. Irrigation is carried out by running the sewage through the fields in trenches and channels. Organic matter in the sewage is broken down by the soil bacteria in the presence of atmospheric oxygen. Crops grown in sewage farms usually produce good yields because of the well-balanced nutrition they receive. But this does not necessarily make it an economic proposition. Sewage farming has never been the major method for dealing with human wastes and its use is now declining very rapidly. Its main disadvantages lie in the requirement of large areas of land to be able to treat the sewage from even a moderately sized town, the unavoidable existence of penetrating smells and the possibility of polluting underground water-courses.

Modern sewage treatment uses all the known techniques of filtration, precipitation and biological activity to produce effluent which is as innocuous as possible. There are certain analytical features that must be looked for during the course and at the end of the treatment. It will be appreciated by the reader that if the treatment is 100 per cent successful in removing all the pollutants and sewage then the final effluent should be pure water and nothing else. In effect the analytical chemist studying the effluent is checking to see how near the quality of pure water it is. It should be added that there have been cases in the U.S.A. during severe droughts in which small isolated communities have had to use purified sewage effluent for drinking purposes with no apparent unpleasant after-effects.

The standards applied to sewage are still based upon those in the eighth report of the Royal Commission on Sewage Disposal published in 1912. The most important test used is a measure of the Biochemical Oxygen Demand (B.O.D.) already described in Chapter 3. The value is expressed in parts per million, i.e. the number of milligrammes of oxygen required to oxidise the organic matter in a litre sample during a period of five days. A low value indicates a correspondingly low organic content, for which the Royal Commission maximum limit is 20 p.p.m. under normal conditions.

The other Royal Commission standard which has to be adhered to is that for maximum permitted suspended solids, for which the level is 30 p.p.m. Samples examined for this purpose are either filtered through a very fine sintered glass plate or centrifuged at high velocity. The residues obtained are then dried at 100–105°C before being left to cool and to be weighed at room temperature. The Royal Commission figures are normally applied in areas in which the river into which the final effluent is discharged leads to a dilution of at least eight times. Where the river flow is too slow for this dilution normally to be attained the standards of the effluent must be correspondingly higher. In many districts, levels much tighter than Royal Commission standards have been imposed. For Cumbernauld New Town in Scotland the B.O.D. figure must be as low as 7 p.p.m. and the suspended solids 10 p.p.m. These are about one-third of the recommended standards and reflect the degree of efficiency that can be expected from a modern sewage purification works.

In many areas additional standards are applied and more extensive chemical tests are carried out to ensure that these standards are attained. There is a whole series of tests to determine the presence of nitrogen in the forms of ammonia, nitrate or nitrite. These are of particular importance if there is any likelihood of the sewage effluent being used for drinking because of the dangers of methemoglobinemia described in the last chapter. Unfortunately there is no generally accepted level for nitrogen in the combined form, although some authorities have set limits as low as 10 p.p.m. for sewage effluent discharged into rivers.

Acidity, measured on the pH scale, is normally adjusted before or at the time any wastes reach the sewage works as the effective functioning of certain stages is dependent upon a fairly neutral pH value being maintained. In many areas the continuous discharge of detergents with domestic waste necessitates the use of a frothing test in order to assess whether or not the treatment of the wastes is successful in breaking down the complex detergent molecules. A volume of the treated effluent is shaken for a set period of time and the time taken for the froth to break down is measured as a guide to the ability of the effluent to foam when discharged.

As the nature of domestic waste becomes more complex owing to the increasing number of chemical compounds being employed in all walks of life, so the number of tests necessary on the treated sewage is also increasing. Fortunately development of analytical techniques and instrumentation have so far managed to keep pace with the problems that have arisen. Because these methods can detect minute quantities of substances they often produce interesting and thought-provoking results. In December 1969 it was announced that one possible result of the extensive use of the contraceptive pill had been the build-up of the level of the female hormone estrogen in the drinking water of the London area. This had arisen because estrogen is quickly eliminated from the human body through the urine and passes with the domestic waste to the sewage works. The treatment process is not 100 per cent successful in removing the hormone so some of it remains in the effluent discharged in the river. As a result of this it passes into the drinking water when water supplies are extracted from the river.

It should in any case be remembered that the human female is always passing this particular hormone in her urine irrespective of whether or not she is on the pill. It has been stated that the levels in any case are so minute that the average person is unlikely to ingest even the equivalent of a single pill during his, or her, lifetime. What nobody bothered to point out was that the method of analysis used for this study was so well developed that it was capable of detecting about one part in one thousand million and that this level of precision can be applied to checking many of the possible pollutants which can be carried in sewage.

For a sewage treatment plant to be really effective it must be a fairly complex place, and sewage engineering is a specialised field in itself. Unfortunately many of the sewage works built thirty or forty years ago are now very much out of date and incapable of dealing with the changing nature of waterborne wastes from enlarged communities.

Modern sewage treatment is based upon biological treatment of the waste which to all intents and purposes is an accelerated form of the bacterial reaction which occurs naturally in the rivers. The first stages of the treatment are intended to create the optimum conditions for the biological processes to take place, in order to obtain an effluent that will ideally require little or no further purification when introduced into the river.

Primary treatment of the sewage involves several steps, of which the first is generally screening. This usually takes one of two main forms. Either screening by itself to remove paper, string, and rags, which might cause blockages if allowed to pass through; or alternatively by passing through a comminutor, which acts as a combined disintegrator and screening device. This is a rotating drum possessing a series of cutters and screens. The screened sewage then passes to what are known as 'detritus tanks', the purpose of which is to enable grit and gravel to settle. This is necessary because the abrasive nature of the particles would create problems with flow control valves and lead to silting throughout the plant. To enable the settling of this material to occur, the detritus tanks are designed to permit a rapid flow of the sewage. By this means the lighter organic matter is carried forward in the flow to the sedimentation tanks.

These are substantially larger than the detritus tanks as they are intended to slow down the velocity of the passage of the sewage. In many works the flow becomes slow enough for it to take up to ten hours for sewage entering at one end to reach the other. During the process of continuous flow sedimentation much of the suspended organic matter settles out on the bottom of the tanks; this is often helped by the addition of special chemical coagulants to the sewage. The effectiveness of this settling stage is of great importance in the treatment process. The greater the degree of settling the lower the strength of the sewage requiring further treatment.

Over a period of time there is a build-up of organic residues on the bottom of the sedimentation tank and these are termed 'sewage sludge'. This is removed by mechanical sweepers which push it to one end of the tank where it can be collected into hoppers fitted into the base of the tanks. At this stage the sludge has a very high water content, usually above 95 per cent. Suction is used to remove the sludge from the bottom of the hoppers and it is further processed in one of several possible ways, as described later. Sedimentation constitutes the completion of the primary treatment of the sewage which still contains both suspended and solubilized matter. Many treatments finish at this point and in these cases the sewage is discharged into the rivers where it adds to the water pollution.

More complete treatment of sewage involves a second stage which constitutes an artificially accelerated form of the natural self-purification processes which occur in the river. Two different forms of secondary treatment are in common use, and the one familiar to most people is the use of percolating filters with their long, slowly rotating arms. These filters are made from coke, slag or similar materials with particle sizes from one-eighth of an inch to several inches in diameter. The filter bed is intended to serve as a medium suitable for the growth and development of a variety of micro-organisms which will break down the organic constituents left in the sewage. Beds vary in size but may be up to 100 feet in diameter and 10 feet deep. The rotating spray distributes the sewage uniformly over the surface of the bed; it filters through and in the process is broken down by the film of bacteria and fungi.

This process is dependent upon an adequate supply of air through the filter bed to enable the aerobic processes to take place. As a result the constituents of the sewage are converted to carbon dioxide, water and inorganic salts. At the same time the B.O.D. of the sewage is progressively reduced so that discharge into a river is not likely to lead to undesirable fish deaths or a lowering of the oxygen level in the water. Before effluent from the filter beds is discharged into a river it is passed through another set of sedimentation tanks known as 'humus tanks'. These are used because the biological processes on the filter beds lead to the formation of brown suspended solids which should be removed by settling out before the effluent is released to the river.

The alternative method of carrying out the secondary treatment is known as the 'Activated Sludge Process', first developed nearly sixty years ago. In this form of treatment the effluent from the primary treatment sedimentation tanks is passed into very large tanks through which it travels slowly for a period of up to ten or twelve hours. During this time it is inoculated with biologically active sludge and continuously aerated by compressed air being pumped in through diffusion units sited in the base of the tanks. Under these conditions very rapid aerobic oxidation of the sewage takes place, similar to that which occurs in the percolating filter beds, and the final effluent is of comparable quality. Before discharge to the river occurs the treated sewage is retained in settling tanks for several hours to enable the sludge to settle. Part of this sludge is returned to the aeration tanks to inoculate further quantities of sewage for the biological treatment.

In cases in which the sewage effluent is required to be superior to the Royal Commission standards it is necessary to treat the sewage beyond the secondary stage. Tertiary treatment of sewage, often known as 'polishing', may take a variety of forms and will become of increasing importance as the demands for more pure effluents increase. In some cases it is sufficient to clarify the effluent by sand filtration or chemical coagulation, as is carried out with water purification. An alternative process is to retain the effluent in lagoons or open ponds for several weeks, during which time further biological reactions occur leading to substantial reductions in any residual impurities. Of major concern now are methods for reducing or eliminating the nitrates and nitrites from the sewage effluent. One method that has met with success is to hold the effluent in a tank growing algae, which extract the nitrogenous materials for nutrition. The procedure works best in warm climates as it is inhibited by low temperatures. As a great deal of the nitrates and nitrites arise from oxidation of ammonia and ammoniacal compounds some approaches to the problem have concentrated on removing these before the conversion has occurred. One particular process for this purpose uses a carefully balanced biochemical reaction with specially treated sewage to break down the nitrogenous compounds and release free nitrogen to the atmosphere.

Whilst the removal of the components of sewage is of major importance, the very procedures adopted in themselves create additional problems. On occasions the purification processes lead to unpleasant odours and there are continuous quantities of grit from the detritus tanks and sludge to be disposed of. Unpleasant odours arise from two main causes: either overloading of the sewage works, in which case the answer is to increase the plant capacity, or contamination and deterioration of one of the purification stages. Both of these are engineering problems and should not be of a permanent nature. Few difficulties occur with the removal of detritus, as its very nature enables it to be used for land-filling and spreading.

Disposal of sludge is of major importance, as it contains a large proportion of organic material which can create the most objectionable odours if disposed of in an haphazard manner. In the interests of civilised society a number of different methods have been devised to use the sludge without it creating a nuisance.

Most methods of disposal necessitate removal of the greater part of the high water content in the sludge as a first step. This is carried out by the standard procedures such as centrifuging, filtration and evaporation. Following this the still wet sludge is in many cases dumped at sea. Both Glasgow and London have special ships for this purpose that can carry large quantities of the sludge away from the coast. Sludge discharged in this way takes about six weeks to be fully digested in the sea, and over a period of time continuous dumping in one area can lead to a growth in localised sea pollution and a corresponding decrease in the amount of dissolved oxygen. For this reason authorities disposing of sludge in this way carry out periodic sampling in the dumping area and if necessary relocate the site if there is any indication of deterioration occurring.

Whilst sea dumping is particularly suited for cities located on estuaries the inland towns must look to alternative methods of disposal. In some places dried sludge is incinerated to produce heat which can be used on the site. Sludge utilisation is of increasing value and importance as the quantities of sewage are increasing every year. Sludge can be, and has been, used as manure for many years. Heat-dried sludge is very common as a fertiliser, and some local authorities offer it for sale to local gardeners as well as to farmers. Raw sludge from the sedimentation tanks is also used after the water content has been lowered. This has the advantage that it can be sprayed straight on to the soil. By this method of disposal the additional expenses involved in drying and packing are avoided. There are, however, many objections to the use of raw sludge in this way because it is likely to contain dangerous bacteria and parasites. Considerable restrictions on its use exist, for example in California where every effort is made to prevent the spread of plant diseases and infections.

For the above reasons most sludge is transferred to digestion tanks in which a natural fermentation process can take place, in the absence of air, by the action of anaerobic bacteria. This process removes any offensive smell from the sludge and leads to the formation of useful waste gases. Fermentation is a slow process taking two to three months if carried out at normal temperatures, so the sludge is heated to just below 35° C to speed up the digestion, which is complete within four weeks at this elevated temperature. Digested sludge from this treatment is much more suitable for direct land spraying than is raw sludge and is extensively used in many places. It has been the standard method of disposal in the Channel Island of Jersey for many years and is supplied continuously throughout the year to the farmers. In 1969 25,000 tons of thickened digested sludge were supplied from the Bellozanne sewage treatment works in Jersey, for spreading on the farms around the island. Demand for this type of manure appears to be increasing and it seems a pity that in other places the sludge is dumped at sea.

Production of sludge gas during the digestion has enabled many sewage works to become self-supporting from an energy point of view. Sludge gas consists of a mixture of methane (70 per cent) and carbon dioxide (30 per cent) with small quantities of other gases. It has a higher calorific value than normal coal gas and resembles natural gas more closely. It is burnt in sewage works power houses either to provide the energy to drive the air compressors for the aeration in the activated sludge tanks or to supply heat to the sludge digestion tanks.

Even with the useful end-products from sewage the complete and adequate treatment of our wastes is not cheap. Running costs for the sewage service in Greater London are about £10 million per annum. A new sewage works in Glasgow or Teesside, dealing with a population running into millions, can cost over £20 million. In Britain we have the knowledge and experience to improve greatly the treatment of sewage. Despite this, many local authorities still give either no treatment at all or only primary treatment, leaving the already heavily polluted rivers to carry out the proper job of cleaning up the human waste.

Even the much-vaunted Greater London Council is still discharging 20 per cent of London's sewage into the Thames after only giving it primary treatment. It will not be until 1973 that improvements at the Beckton sewage works will enable full primary and secondary treatment to be given to all the waste arriving there. In some areas local authorities turn a blind eye to the pollution caused in rivers by their own sewage. If it was not so pathetic it would be laughable to read of the chairman of the magistrates at Gosforth reprimanding a ship's captain, found guilty of oil pollution, with the words 'It was unfortunate accident but we are concerned over pollution of the River Tyne' – whilst a few miles up-stream Newcastle is pumping millions of gallons of untreated sewage into the river every day.

One of the major limitations of many sewage works is that water from street drainage is often combined with the flow of sewage and water-borne wastes. Under dry weather conditions this has no effect on the sewage works, but in wet weather the treatment plant has to cope with abnormally high loads in which the sewage has been considerably diluted. The drainage water contains very little in the way of pollutants and ideally could be allowed to drain straight into the rivers. Such an idealised situation can only occur in areas in which the road drainage pipes are separate from the sewage pipes. Where surface drainage water runs with the main sewers and dangers of overloading at the sewage works exist it is normal to have overflow pits to enable excess loads to flow straight to the outfall pipe into the river. Inevitably this means that some untreated, or only partially treated, sewage gets carried into the rivers. Any solution to this problem will have to be a very long-term one, as it necessitates construction of entirely separate drainage systems in all towns for the surface water.

Although sewage treatment plants vary enormously as to the efficiency with which the different part of the plant operate, the following figures give an idea of the improvement that is to be expected:

	Initial level taken as 100%	% Remaining after primary treatment	% Remaining after primary plus secondary treatment
B.O.D.	100	65	10
Suspended solids	100	50	10
Bacteria	100	50	5
Nitrogen	100	80	50
Phosphates	100	90	70

From these figures it can be seen that failure to give secondary treatment to the sewage leads to substantial quantities of pollutants still remaining in the effluents discharged into the rivers. As the B.O.D. in strong sewage can be as high as 600 p.p.m., primary treatment in isolation produces results that are nowhere near the Royal Commission standards. It should at the same time be pointed out that the cost of sewage treatment is roughly doubled by introducing a secondary treatment stage. Incorporation of extra processes to 'polish' the sewage and remove greater proportions of phosphates and nitrogenous materials increases the cost even further.

When it is considered that sewage treatment for a population of fifty thousand people and its attendant industry necessitates dealing with about three million gallons per day distributed in a non-uniform manner throughout the twenty-four hours it will be appreciated that full treatment cannot be achieved cheaply. It has been estimated that a new sewage works costs up to £15 per head for the population it serves and the running costs can be as high as £2 per head per annum. Such costs are increasing rapidly because of rises in salaries and the costs of replacement machinery. These costs are only partially offset by revenue from dealing with trade wastes and the sale of manure. Many people feel that even these costs are better than seeing untreated human excreta hanging from outfalls above our rivers and preferable to the dangers of fouled beaches. The experiences of polluted beaches during the British 'Dirty Jobs' strike in the autumn of 1970 was enough to help people appreciate the true value of full sewage treatment.

Fish deaths, such as this observed in a canal in Florida, U.S.A., can occur either from pollutants in the water or from lack of dissolved oxygen. The return of aquatic life to a river is evidence of a decrease in these forms of pollution.

The battle against pollution is not a new one. The Water Pollution Research Laboratory at Stevenage, England, was first established in 1927 and has pioneered many of the procedures for sewage treatment and water purification.

Signs such as this were displayed around the coast of Britain during European Conservation Year 1970 because of the 'dirty jobs' strike of sewage workers and refuse collectors. As a result pollution became a talking point almost overnight.

Because of the inadequacies of municipal waste disposal services a large part of industry has to hire private contractors to deal with industrial wastes. Tankers carrying wastes, such as this belonging to the Purle Group, are now common sights on British roads.

Disposal of industrial wastes takes many forms, desirable and undesirable. In Co. Durham colliery wastes are continuously tipped into the sea – to be washed up on the beaches further along the coast.

The busiest sea route in the world is the English Channel, and collisions between ships are frequent. This blaze on the oil tanker *Pacific Glory* occurred after it had been in collision with the *Allegro*. Fortunately only a small quantity of the 70,000 tons of crude oil escaped.

The French and British coasts were far less fortunate with the *Torrey Canyon* disaster, and large quantities of crude oil which escaped from the wreck were burnt on the beaches.

6 The air we breathe

Unlike water pollution, which for the greater part affects the nations causing the pollution, air contaminants become dispersed around the world with the result that even underdeveloped areas receive the fall-out from industrialised countries. This is evident from the progressive spread of radioactive particles resulting from the explosion of nuclear devices in different parts of the globe. Dispersion and dilution, however, do take time, so that temporary localised concentrations persist, depending upon wind and weather conditions. The United Kingdom is very fortunate in that it is a collection of islands with frequent rainstorms and good winds. These combined factors help to clear away the numerous pollutants for which we are all responsible. At the same time we must not, however, become complacent or any less conscious of the continual build-up of pollution that is occurring on an international scale.

In Great Britain seriously polluted atmospheres have existed for at least seven hundred years. The first protests about the use of coal as a fuel were made in the reign of Edward I. Despite the increase of smoke in major cities such as London successive monarchs did very little, though Queen Elizabeth I did prohibit the burning of coal while Parliament was sitting. Parliament in its turn procrastinated for years by setting up commissions and studies and taking no action. By the reign of Charles II the situation had become so bad that John Evelyn, in his *Fumifugium,* wrote of London: 'Her inhabitants breathe nothing but an impure and thick mist, accompanied with a fulginous and filthy vapour, which renders them obnoxious to a thousand inconveniences, corrupting the Lungs, and disordering the entire habit of their bodies; so that Catharrs, Phthisicks, Coughs and Consumptions, rage more in this one City, than in the whole Earth besides.'

He recommended that all industries forming smoke should be moved five or six miles away from the City, and that there should be restrictions on further residential development as well as increased planting of trees and shrubs. Little or nothing was done to put these suggestions into practice and the situation in London and other major cities continued to deteriorate, particularly when the industrial revolution got under way. This led eventually to the 1863 Alkali Act which, with various modifications, was virtually the only control on any form of atmospheric pollution that existed until the 1936 Public Health Act. The first big step forward in the field of restriction and control did not come, however, until the 1956 Clean Air Act was implemented as a result of the Beaver Committee Report on the 1952 London smog.

It has been traditional for foreigners to believe that British cities are permanently shrouded in fog and that every winter sees the descent of a thick cloud of smog which does not lift until the spring. Anyone who has encountered the air pollution and eye irritation in cities like New York and Los Angeles will know that, for all their disadvantages, Manchester and London are far preferable places in which to breathe. Britain's reputation for fog has grown, and been perpetuated, by the popularity of novelists such as Charles Dickens and by films usually showing murders carried out in London fogs. A more accurate indication of what has actually been achieved in Britain was seen during the summer of 1970, when many of the major cities in the world were facing crisis situations over pollution, at a time when Britain was enjoying the sunshine with no adverse effects from photochemical smog.

Air pollution, particularly in cities, is a world-wide problem and most of the answers in one part of the world are equally applicable elsewhere. The big London smog of 1952 served to focus attention on what could result from a combination of fog, smoke, vehicle exhausts and low temperatures all occurring at the same time as a temperature inversion. The situation in fact applied to most of the British Isles, but was particularly acute in London. By the second and third days of the 1952 smog hospital admissions for respiratory disorders had climbed rapidly and remained above average for several weeks. Although the smog itself cleared on the fifth day, studies of hospital admissions indicated that over an extended period it was probably responsible for 4,000 deaths. This was not the first time, nor the last, that something of this nature arose either in England or in the world, although it was probably the worst. A similar situation occurring in London in 1956 led to 1,000 deaths.

One of the greatest contributing factors to localised smog situations is the existence of temperature inversions. These occur when a layer of cold air is trapped under a layer of warm air and unable to move because of surrounding hills or mountains. This then acts as a trap for smoke and fumes which get held within the area covered by the inversion. The situation is particularly aggravated when the sun acts photochemically on the pollutants, as in Los Angeles, and causes the formation of chemical irritants which are harmful to both eyes and lungs.

An inversion of this nature was the cause of at least 60 deaths in the Meuse valley in Belgium in 1930, when for nearly a whole week the pollution from a highly industrialised area was trapped, and produced the widespread respiratory disorders which are now known to be so common in this type of situation. Many instances are known of similar situations occurring in the U.S.A. One particularly bad case was in Pennsylvania in 1948 when the town of Donora suffered from this fatal combination of fog and temperature inversion. Whilst the number of deaths directly attributed to the smog was only about 20, it was estimated that over 40 per cent of the population of 14,000 suffered from some form of ill-health.

It should be emphasised that whilst the overall health effect is essentially the same, the smog situation created from a combination of fog and industrial pollution during damp, cold weather, as in the above cases, differs from the photochemical smogs occurring during hot weather. This second type of smog was the cause of international concern during the last week of July 1970, when many of the major cities including Tokyo, New York, Los Angeles, Milan and Buenos Aires were simultaneously affected. The combination of hot weather and pollution virtually brought New York to a standstill, whilst 9,000 people were reported to be sick from the effects in Tokyo.

Because it is always around without being seen there is a tendency for people to take the air for granted, and to assume that it can absorb unlimited quantities of smoke and chemicals without anything happening. Clean, dry air is a complex mixtures of gases and when dry contains 78·9 per cent nitrogen, 20·94 per cent oxygen and 0·97 per cent of other gases. Apart from a number of unreactive rare gases, equal to 0·93 per cent, this small residual amount includes carbon dioxide which constitutes 318 parts per million (0·0318 per cent in the air), plus methane, hydrogen, sulphur dioxide and various nitrogen compounds totalling 2·5 p.p.m. Under normal atmospheric conditions moisture in the atmosphere varies between 1 and 3 per cent.

The major pollutants released into the atmosphere are carbon monoxide, sulphur dioxide and sulphur trioxide, nitrogen oxides, hydrocarbons and sundry particles. There is hardly a human activity which does not at some stage add to the air pollution. Fortunately, like the rivers, the air does have a certain ability to clean itself. Particulate matter is slowly deposited over large areas often many miles from its source, hydrocarbons eventually break down into more simple compounds and settle out, and the various acidic gases are dissolved by rain and get carried on to the land. Even carbon dioxide is removed from the atmosphere by plants and solution in the sea. The concern which exists really covers two separate problems. First there is the problem of localised concentrations of pollution which are a danger to health and property, and secondly the long-term effects of progressive build-up of atmospheric pollution at a faster rate than can be coped with by the self-cleaning mechanism of the air.

Combustion of fossil fuels, such as coal and oil; incineration of garden refuse; and exhaust gases from vehicles, give rise to gases containing carbon. Thus carbon dioxide, carbon monoxide and hydrocarbons are all products of the internal combustion engine. Localised concentrations of carbon monoxide frequently exceed 10 p.p.m. in highly industrialised, motorised areas, and on occasion levels exceeding 100 p.p.m. have been measured. Carbon monoxide is a colourless, odourless gas, and its extreme toxicity means that concentrations that build up in traffic jams are quite enough to give car travellers headaches or a feeling of sickness. It is not yet clear how carbon monoxide is removed from the atmosphere, although it seems certain at present that there is no measurable increase in the minute global level existing despite the millions of tons released into the atmosphere every year. It is quite likely that it is oxidised photochemically to carbon dioxide.

Despite their similarity in names, carbon dioxide is chemically very different from carbon monoxide. It is not usually considered to be a pollutant as it is a natural constituent of air. It has been claimed, however, that increases in the carbon dioxide content of the atmosphere arising from man's combustion processes have been responsible for changes in the world's climate. Predictions have been made that continued increases could lead to the melting of glaciers and icebergs and the drowning of many of the major cities. The basis behind this reasoning is that carbon dioxide in the air traps invisible infrared radiation from the sun, which would be reflected from the earth's surface in the absence of the atmosphere. The trapped radiation absorbed by the carbon dioxide, instead of escaping to outer space, thus contributes to maintaining the atmospheric temperature. When an increase in carbon dioxide content occurs, the quantity of radiation that can be absorbed also increases, leading to an overall increase in atmospheric temperatures.

Various very careful studies show that global concentration of carbon dioxide has risen from 290 p.p.m. at the end of the nineteenth century to about 318 p.p.m. by 1970. This is a total increase of about 9·5 per cent and represents average annual increase of roughly 0·4 p.p.m. Rises have probably been higher in the past thirty years than previously, and statistical studies suggest that the actual increase is currently fairly steady at about 0·7 p.p.m. per annum. Conflicting figures exist concerning the extent that the world temperature will rise relative to an increase in the carbon dioxide content. One authority claims that a 10 per cent increase in carbon dioxide produces a 0·5° C increase in temperature, whereas another considers that a 30 per cent increase would be necessary to give this temperature rise.

The whole discussion is further confused by the observation that whilst world temperatures were increasing until about 1940, after that date they started to decline. It has been suggested that this decline has arisen from the increasing turbidity of the atmosphere from pollution preventing the sun's rays from penetrating in the first case, thus counterbalancing any effect of increased carbon dioxide. It does seem that the world is not yet in danger of drowning, although the present rate at which the carbon dioxide level is increasing suggests that global temperatures could rise by a further 1 to 2° C by the end of this century if the pre-1940 increase again prevails. When it is considered that there is fifty times as much carbon dioxide in the sea as in the atmosphere and the complete carbon cycle of interchange between air, water, plants, animals and decomposition involves millions of tons of carbon dioxide every year, it is difficult to make any real sense from the greatly conflicting evidence.

One point that stands out above everything else is that man is using up the fossil fuels at a tremendous rate, and this is increasing the carbon dioxide content of the atmosphere. From a public health point of view this does not constitute any form of a hazard, and there is not yet sufficient knowledge to be able to claim that it has had a material effect on the earth's weather. There is in any case an indication that the water vapour, formed from the combustion at the same time as the carbon dioxide, has an opposite effect on the sun's rays and so counteracts any rise in temperature. Increases in dust particles and sulphur dioxide also add to what has been called the 'dirty greenhouse' effect and counteract any global increase in temperature due from carbon dioxide. Current knowledge is inadequate for there to be any real justification for the prognostications from the prophets of doom.

Nitrogen oxides that are formed from combustion processes

are nitric oxide and nitrogen dioxide. These are active constituents in the photochemical formation of eye and lung irritants. Because of their acidic character they dissolve easily in water and are fairly rapidly removed from the atmosphere by rain and other chemical reactions in biological respiration and food cycles. No substantial increase in atmospheric levels therefore occurs.

Sulphur dioxide is formed either from direct combustion of fossil fuels containing sulphur impurities, or from atmospheric oxidation of hydrogen sulphide, which is also released by decaying matter and some industrial processes. Sulphur dioxide is itself oxidised in the atmosphere to sulphur trioxide, which forms sulphuric acid when dissolved in water. This combination of oxides and acid which can reach significant levels in wintry foggy conditions can lead to serious health problems, as the danger level for sulphur dioxide can be a concentration of as little as 1 p.p.m. During the 1952 London smog an average figure of 1.34 p.p.m. was obtained over a period of two days. Sulphur that passes into the atmosphere in this form is returned to the earth within about two weeks, most of it being carried away during rainfall. It plays a fairly important role in the environmental sulphur cycle which involves plant incorporation and solution into the sea. All of these processes would continue quite happily without the added quantities discharged by man. It is the localised concentrations of sulphur dioxide and its derivatives which can and do cause damage.

Because of the large quantities of fuel that they use, power stations are responsible for emitting large amounts of sulphur dioxide. If allowed to disperse with no special treatment, the gas rises to a high altitude and gives rise to no apparent localised concentrations. If the gas emissions are treated chemically, 90 per cent of the sulphur dioxide can be removed, but the residual gases become cooled and dense so that very poor dispersion takes place. This state of affairs led to the 'washing' of gases from Battersea Power Station, London, being suspended in 1970 for a two-year experimental period as people in the immediate vicinity were suffering from the remaining 10 per cent of sulphur dioxide.

Apart from respiratory harm to humans and animals, even concentrations of sulphur dioxide in the atmosphere well below 1 p.p.m. cause plant damage. This occurs because the gas is absorbed into the cells of the leaves of the plant and the acidic nature causes a breakdown in the green chlorophyll colouring matter that is so important for the plant's respiration. During the past ten years there has been a localised rise in the sulphur dioxide content of the air over Europe leading to a corresponding increase in the acidity of the rainfall. This has provoked major concern for the pine forests in Sweden, where it has been claimed that plant growth is being stunted and that lakes are becoming more acidic and likely to become unable to support fish life. This is now an international issue, although the American Chemical Society in their publication *Cleaning Our Environment* have pointed out that man-made emissions of sulphur compounds are estimated to account for only one-third of the total amount entering the atmosphere each year, the remainder coming from natural sources. The main reason that steps need to be taken to reduce sulphur dioxide levels is because of the dangers to health arising from localised concentrations.

Another constituent of air pollution that is creating a certain amount of concern is the existence of the chemical hydrogen fluoride, which usually arises from industrial processes such as brickworks and some fertiliser plants. Biological investigations have shown that although atmospheric fluorides rarely attain concentrations greater than one-fiftieth of 1 p.p.m. they act as a cumulative poison in plants. That is, the plant appears to absorb the fluoride from the atmosphere as a continuous process but does not release it at the same rate, with the result that there is a progressive increase in the plant fluoride level. Eventually this reaches a toxic level and leads to a collapse of the cellular structure in the leaves which is visibly observed as a brown, dead margin around the leaf. Factories producing phosphates are among the worst emitters of fluoride, and in one area alone in Florida, U.S.A., are responsible for the release of seven tons a day. This emission has been credited with being responsible for killing plants and animals in addition to creating severe respiratory disorders in humans and almost intolerable living conditions.

Photochemical reactions leading to the formation of the Los Angeles type of smog, which does not involve smoke or fog, have been under investigation since the serious nature of this form of air pollution was first recognised as arising from a combination of strong sunlight and vehicle fumes. In actual fact man is only responsible for about 15 per cent of the total emissions of hydrocarbons into the atmosphere. Natural sources of these chemicals include trees and other plants which release terpenes, plus the bacterial decomposition of matter which forms methane. Once again the disadvantage of man's contribution lies in the fact that his pollutants are concentrated into limited areas in which the maximum harm is created. The combination of hydrocarbons, nitrogen oxides and a supply of oxygen in the presence of plenty of sunshine is enough to lead to a series of large-scale reactions taking place. The products of these reactions are both unpleasant and injurious to all animal life, so that when concentrations build up, living conditions in such areas can become intolerable. It should be borne in mind that such a situation can occur in any urban area if the correct combination of factors exists and is not necessarily restricted to cities in the U.S.A.

The full nature of the reaction creating photochemical smog is not yet certain, but it involves the formation of active chemical intermediates from the hydrocarbons known as free radicals. Roughly the chain of reactions taking place can be shown to be:

1 Sunlight acting on nitrogen dioxide produces nitric oxide plus atomic oxygen;

2 Atomic oxygen plus atmospheric oxygen produce ozone (a very active form of oxygen);

3 Ozone plus hydrocarbon produce oxyhydrocarbon radicals;

4 Oxyhydrocarbon radicals break down to give aldehydes and ketones;

5 Oxyhydrocarbon radicals plus nitrogen dioxide produce peroxyacylnitrates.

From this representation it can be seen that to reduce or eliminate the occurrence of photochemical smog it is necessary to prevent the formation of nitrogen oxides and the emission of hydrocarbons into the atmosphere.

It often surprises people to learn that lead is a common air pollutant and this turns to astonishment when they are told that it arises from the lead additives in petrol. Lead (in the form of lead tetraethyl) is used to increase the octane rating of the petrol in order to make it suitable for high compression engines. Most of the lead additive finishes up being discharged from the exhaust. Because of this lead emissions are concentrated along the highways, and have been detected as deposits on plants one hundred and fifty yards from the roadway. Most petrols contain between 2 and 3 grammes of lead in every gallon and conversion to lead-free petrol would probably mean increased transportation costs. However, lead emissions are not as dangerous as some of the others, and they are mainly removed during the washing of vegetables anyway. Lead is a very common element and occurs naturally in many foods. It has been calculated in the U.S.A. that an adult ingests 0·3 milligrammes of lead every day, but even in cities only about 17 per cent of this comes from the atmosphere.

In addition to the gaseous chemical nature of air pollution there is also a very wide range of particles which are responsible for different pollution effects. Some of these particles are simply liquid droplets which have condensed around microscopic bits of dust. Others are solid residues from industrial smoke-stacks or particles of earth blown around by the wind. Fogs are made up from fine moisture droplets, 'aerosols', which have condensed around solid particles, and unless wind or rain lead to dispersion of the fog the droplets act as traps for smoke and other pollutants. Even the ubiquitous motor car is once again a culprit here, as the disintegration of tyres and wearing of the road surface both add to the number of fine particles in the atmosphere.

Because of the varying nature of particles it is not possible to deal specifically with each form, as sizes range from less than 0·1 micrometre in diameter to above 10 micrometres. Most particles are unlikely to remain in the atmosphere for more than two or three days as they will be removed by rain, settling, and by impact on stationary objects. Large particles tend to settle quicker than fine particles, and the variation in size leads to settling out over a wide range of distances from the source of the emission. The presence of large concentrations of particles in the atmosphere contributes to haziness and reduces the amount of sunlight falling on the earth. Quantities of solid matter settling out in heavily industrialised areas often exceed one ton per square mile each day, and ocasionally may be three or four times this level. This corresponds to more than 40 lb of rubbish settling in the average back garden every year.

While they are in the atmosphere the particles are constantly undergoing changes in size and shape. Their chemical composition also changes as they coalesce and combine with each other, as well as absorbing gases and vapours from the surrounding atmosphere. In the study of atmospheric particles it is now common to determine the proportion that will dissolve in a sample of benzene (this is benzene-soluble organic material), and the amount left as insoluble (mainly inorganic) solids. Some of the organic constituents are complex substances known as polynuclear hydrocarbons. A number of these are known to be carcinogenic – they have been observed to cause cancer growths in laboratory investigations on rats. Such compounds arise from pyrolysis (subjection to high temperatures) of organic material and are commonly associated with smoky fall-out.

Investigation of air pollution and studies on the nature of pollutants and fall-out have been carried out for many years. The Warren Springs Laboratory of the Department of the Environment at Stevenage is responsible for a National Survey of Air Pollution in the United Kingdom. It correlates data from 1,000 sampling sites in 200 towns, which operate a 24-hour measurement of smoke and sulphur dioxide. Figures obtained from this are used as the basis for showing the effectiveness of legislation brought in to control pollution and serve to indicate if there is any new source of emission in any particular area. Most other industrial countries now conduct similar surveys. The corresponding survey in the U.S.A. is carried out by a National Air Sampling Network and the results are co-ordinated at the National Center for Air Pollution Control. It is not always possible to relate figures obtained in one country with those in another because of differences in sampling procedures and methods of analysis. Because of this the Organisation for Economic Cooperation and Development published recommended methods for measuring most of the common pollutants and included international standard charts to be used in the study of suspended particles. As more and more countries are becoming concerned about air pollution it is increasingly important that international standards are applied.

Sampling methods employed in pollution studies depend upon the nature of the substance under investigation, which may be a gas, an aerosol or solid particles. In most cases the investigator will be concerned with obtaining as large a sample as possible in the minimum period of time, and for this purpose will use a pump to suck the air through a collector at the sample site with a meter to show the volume of air passed. On a national scale sampling sites are selected in open country as well as near to obvious sources of pollution, or where temperature inversions occur. In the Greater London area alone sampling is carried out at twenty-seven sites for such things as deposited solids and sulphur dioxide content.

Individual substances are obtained by sucking the air through a suitable collecting medium which will filter it out and, if possible, isolate it from the other air pollutants. Thus sulphur dioxide can be absorbed by passing it through water containing a small quantity of hydrogen peroxide. This leads to the formation of sulphuric acid in the aqueous solution which can be measured by standard chemical methods of titration and also colorimetrically. Similarly hydrogen sulphide is studied by absorbing it in a solution of cadmium hydroxide; leading to the formation of a corresponding amount of cadmium sulphide, which is measured by the intensity of a colour reaction it will undergo with *p*-aminodimethylaniline. The nitrogen oxides which are so important in the studies of photochemical smog are also measured by a colour reaction that they will undergo. In this case the reagents are sulphanilic acid and α-naphthylamine which, when combined in the presence of the nitrogen oxides, give a pink-coloured compound.

Many methods exist for the determination of the ubiquitous carbon monoxide and are extensively used at different sample sites. One of the traditional methods involves passing the air containing the carbon monoxide through a tube packed with iodine pentoxide. Any carbon monoxide is oxidised to carbon dioxide and in the process releases iodine. Both the quantity of carbon dioxide and the quantity of iodine correspond to the amount of carbon monoxide that has passed through. Normally it is the latter which is then measured chemically and the original concentration of carbon monoxide found. As the number of studies has increased rapidly over the past few years, more automatic methods for carbon monoxide studies have been introduced. One procedure has been the use of gas chromatography in which the carbon monoxide is isolated from the other constituents by passing the air through a long column packed with chemicals which separate the components according to their molecular properties. When the carbon monoxide leaves the column, it is converted by a chemical reaction into methane. This is measured by burning it in a special flame ionisation detector giving an electrical signal with a magnitude corresponding to the amount of methane burnt, and hence to the original amount of carbon monoxide that was present in the sample.

Collection of aerosol samples can pose a number of problems as the investigator will be trying to obtain liquid droplets without at the same time collecting particles or increasing the level of dissolved gases. Aerosols can be collected by sucking the air through very fine pre-weighed filters made from paper or asbestos. The quantity of aerosol is then obtained by reweighing the filter after the sampling has ceased. This type of determination can, however, also pick up solid particles which have to be removed from the face of the filter pad, and other methods have had to be devised to overcome this in cases in which these particles create serious problems. In one of these procedures the air intake is directed on to a number of microscope slides by a series of jets of decreasing orifice sizes. The size of the aerosol droplets collected decreases with the jet sizes.

Because of the increase in hydrocarbons and other complex organic chemicals polluting the atmosphere during the past 20 years there has been a great deal of research into methods for measuring these, which can be applied on a general basis. Many of the techniques developed necessitate the initial freezing out of the organic substances by passing the air through a flask contained in a low-temperature trap using liquid nitrogen or liquid air. By this means the sample obtained can be investigated by more sophisticated laboratory procedures which will give accurate results with minute quantities of materials.

It is outside the scope of this book to describe the operation of such instruments as mass spectrometers, and infrared spectrophotometers, which have been used to study hydrocarbon emissions from refineries and vehicle exhausts. These instruments can differentiate between the various organic chemical compounds on the basis of the individual spectra which they each give. A spectrum of any individual compound is characteristic for that compound under specified conditions and serves as a 'fingerprint' for identification purposes. The methods applied to organic compounds can rarely be carried out at the sampling site, which necessitates taking the sample back to a laboratory suitably equipped with the very expensive apparatus used in these determinations.

Solid particles in the atmosphere are collected by two main methods, both of which are of value for particular purposes. The first method is that of fall-out settlement, in which the particles are collected by settling on plates or in jars exposed to the atmosphere. This is usually carried out over an extended period of several days or even weeks. It does not give a representative picture of the particles in the atmosphere in the sampling area, but only of those settling out. Its special value is therefore in cases in which residents complain about particles from a smoke-stack falling on their washing or upon babies in prams. A particular disadvantage of this method of study is that the sample does not correspond to a definite volume of air and it is only possible to express results as so many grammes per day per square metre. The alternative method of sampling involves sucking measured volumes of air through special membrane filters which are designed to retain the particles on the surface of the filter. This means that after collection the particles can be easily brushed off the membrane for microscopic and chemical studies to be carried out.

Solid samples obtained by such procedures are examined by a variety of methods, of which the first is usually counting. For this purpose the particles are suspended in a suitable liquid and counted visually under a microscope. From the number of particles in a particular volume of the solution it is possible to calculate the number of particles per cubic foot of air originally sampled. Microscopic study can also be used to obtain some idea of the sizes of the particles obtained, and an experienced microscopist can often gain a fairly good idea of the physical and chemical nature of many of the components. There are obvious differences between fibrous particles, plant seeds, combustion particles and industrial fallout, and many of these are particularly noticeable under special illumination which can assist identification.

Chemical investigation of particulate matter falls into two fields: investigation for inorganic material such as fine particles of cement or iron ore, and investigation for organic pollutants such as pesticides and polynuclear hydrocarbons.

An early indication of the inorganic nature of the particles is obtained in the microscopic study of the material and gives a guide to the future route of the investigation. This will involve chemical reactions to check the characteristics of the suspected substance finishing with a spectroscopic check of its 'fingerprint' to confirm the identification. In the case of organic materials complete identification is often very tedious and time-consuming, as there are millions of different organic compounds, and even those used as pesticides and produced in combustion run into thousands. The path of the investigation in this case first involves extracting all the organic matter from the solid particles with a solvent such as hexane or benzene. This not only removes solid organic materials but also any organic liquids which have become adsorbed on to the trapped particles. Evaporation of the solvent leaves the organic residue which has to pass through an involved process of solution, extraction, precipitation and separation with an established series of reagents, just in order to classify any individual component into the chemical group to which it belongs. Complete identification is only achieved after a pure sample of any organic sample has been obtained and subjected to the spectroscopic 'fingerprinting' techniques already mentioned. This very involved procedure is particularly necessary for the correct classification of polynuclear hydrocarbons when the scientist may be checking to find if a combustion source is giving rise to dangerous carcinogenics. An incorrect assignment of chemical structure in a case like this could be very misleading, so it is imperative that the best possible methods are used.

Air pollution controls and abatement have been of growing importance because an increasing number of people have become concerned over the effect it has on their health. Effects on health should not only be measured in terms of the estimated increased number of deaths occurring because of some particular fallout, but also in terms of the numbers of people made sick, the increased number of visits to doctors' surgeries and the increased number of people taking a day off from work. Air pollution does not only kill, it also debilitates, and it is now well known that some people are more susceptible to particular forms of pollution than are others. Elderly people are always more prone to respiratory disorders than younger people, and suffer that much more during periods of heavy pollution.

Apart from eye irritation, which can affect many people severely, the main result common to most forms of air pollution is aggravation of bronchitis and/or the development of emphysema. Both of these are respiratory disorders, the former involving a persistent cough and continuous production of sputum, the latter being the name for un-uniformly distended air sacs in the lungs. In Great Britain chronic bronchitis is the fourth major cause of death and is referred to abroad as 'the English Disease'. Whilst air pollution can in no way be credited with the entire blame for these disorders, it can certainly be shown that continual exposure to many pollutants leads to damage of the lung tissues and progressive deterioration of the mechanism of breathing. It should be noted that cigarette smoking is a localised concentration of air pollution for a short period of time and can be held responsible for some contribution to the present figures for respiratory disorders. Medical investigations continually show a direct relationship between air pollution and illness.

One particularly extensive survey spread over several years has been carried out by Professor P. J. Lawther. He has studied the daily fluctuations of the health of patients suffering from chronic bronchitis and related them to the daily figures for smoke and sulphur dioxide. The research has shown quite clearly that the level of illness increases in close relationship with both the pollutants measured. It is very difficult to obtain any clear comparison of respiratory illnesses between different countries because of variability in diagnosis and different methods of classification. Attempts to standardise on these have been made and future comparisons may be more accurate than past ones have been.

Inhaled air pollutants can create physical damage to the body in a number of different ways. A gas such as carbon monoxide interferes with respiration directly by being preferentially adsorbed by the blood, and if present in a large enough quantity can lead to death. Sulphur dioxide on the other hand dissolves in the lung fluid to form an acidic solution which can destroy the lung tissue. Solid particles may damage lung tissue by abrasion, making it easier for infection to occur, or alternatively may be dissolved by body fluids and lead to poisoning in some other part of the body. There is now considerable evidence to show that pollutants can be linked with a wide variety of disorders mainly connected with the respiratory tract.

Eye irritation is by far the most common discomfort linked with photochemical smog and a major cause of complaint in industrial towns. The onset of eye irritation frequently serves as an indication of the build-up of pollution and can act as a warning to bronchitis sufferers to take precautions. In Los Angeles eye irritation occurs for about two-thirds of the days in the year.

Estimates of the cost of air pollution run to fantastic figures. The Beaver Committee, set up after the 1952 London smog, calculated that at that time the combined total of cleaning, painting, corrosion, loss of efficiency and fuel wastage was costing Great Britain at least £300 million every year. Comparable figures of between £6 and £10 per head per annum have been made in the U.S.A., although estimates as high as £20 per head per annum have been obtained for heavily polluted areas. The cost of controlling the pollution is also astronomical, as the situation has been allowed to deteriorate for such a long time.

Any estimates for the expense involved in reversing the overall upward trend of air pollution are purely speculative, but could well be within the region of £7 to £10 per head per annum over a period of thirty years. This cost would automatically be partially offset by a reduction in the expense of clearing up the ravages of pollution. It should be pointed out that these calculations are hard financial figures relating to improving industrial plant and overcoming domestic emissions. They do not allow for human suffering and the shortening of a person's life. It really becomes a question of what value we put on being able to live a longer life in a cleaner, fresher environment. Simple economics has ceased to be the only criterion for assessing such matters.

During the past five years laws against air pollution have been rapidly tightened up in many countries. It is, however, not uncommon to find overall responsibility for clean air divided between several different bodies. The legal position in the U.S.A. was a very difficult one for many years, because although the Federal Government could legislate on interstate matters and make strong recommendations for action, the major legislation still had to come from the fifty individual State Legislative bodies. This inevitably meant varying degrees of control throughout the country with some areas dragging their heels. Authorities were often afraid that tight controls might drive industry to cities in which a more lax attitude prevailed. The situation was partially resolved in 1963 by the Federal Clean Air Act and further strengthened in 1967 by the Air Quality Act. These two Acts give the U.S. Secretary of Health and Welfare considerable power to advise and financially aid pollution control. However, it still leaves the primary responsibility in the hands of the States.

A similar division of responsibility exists in Great Britain because of the peculiar manner in which control of air pollution has developed. In practice this is covered by two separate sets of legislation which started with the Alkali Act of 1863. The purpose of this Act was to control the discharge of hydrochloric acid into the atmosphere which occurred during the manufacture of sodium sulphate. Under later extensions to the Act between 1906 and 1966 the control of emissions from many other factories came under the scrutiny of the Alkali Inspectorate now responsible to the Department of the Environment. The Alkali Inspectorate covers a wide range of industrial processes including electricity power stations, gas and coke production, smelting works and acid manufacturing. The Act imposes controls requiring owners of such businesses to use the best practicable means to prevent 'noxious or offensive gases' from escaping. No authority has been given to the Alkali Inspectorate to deal with domestic air pollution which is normally in the form of smoke emission. The first attempt to take steps to control industrial smoke emission was under the 1936 Public Health Act, which required that sanctions could be taken in cases in which the emission of smoke could be shown to be a nuisance. This did not apply to private residences despite the fact that at that time they were responsible for producing nearly half the smoke in the atmosphere.

No effective national control of domestic air pollution occurred until the 1956 Clean Air Act was introduced, leading to the creation of smokeless zones, although provision for the creation of such areas was included in the Manchester Corporation Act of 1946. Responsibility for the implementation of the 1956 Act and creation of the smokeless zones was vested in the local authorities, but they had no power to impose controls upon the industries covered by the Alkali Inspectorate (who had their powers extended to include smoke control in these cases). This gave rise to such anomalous situations as local authorities creating smokeless zones for domestic households, while factory chimneys in the same neighbourhood continued to produce smoke. In fact, the 1956 Clean Air Act permitted the emission of dark smoke for certain well-defined periods during the day for industrial purposes, and this aspect of the Act was criticised for its apparent leniency. Considerable strengthening came about as a result of the 1968 Smoke Control Act.

The situation as it now exists is that it is an offence for the domestic householder to emit smoke or burn other than approved fuels in a smokeless zone. At the same time industrial premises are still permitted to produce dark smoke for limited periods. Continuous production of smoke is restricted to four minutes in any eight-hour period, with a maximum total of fourteen minutes being allowed in any eight-hour period for soot-blowing. That these limits are frequently ignored is only too apparent to people who have timed the emissions from smoke-stacks.

Measurement of the intensity of smoke (which is generally considered to be only the black soot particles in combustion gases) is normally done by comparison with standard cards known as Ringelmann Charts. These cards measure 26 inches by 10½ inches and are divided into squares by horizontal and vertical black lines. The thickness of the lines increases through the series of cards – the first card being blank is designated Ringelmann 0 and the area covered by lines increases by 20 per cent through the cards, so that Ringelmann 5 is completely black. The cards are placed in line with the smoke-stack and viewed from 50 feet away. By this means the smoke is given a Ringelmann number corresponding to the chart it most nearly matches. For most practical purposes, on the spot readings are taken with smaller, more portable charts. Within the smoke stack itself control of emission and measure of the smoke intensity is carried out with optical instruments specially designed for this purpose. Unfortunately charts and instruments are frequently inadequate for measuring emissions of particles and gases which are brown, red, grey, white or anything but black. This limits the effective control on many industrial processes.

Further reduction in air pollution can be brought about in several ways. An increase in the development of smokeless zones will lower the amount of domestic pollution by making householders use clean burning fuels such as anthracite and coke. Even these, however, will still produce large quantities of colourless sulphur dioxide and reduction of this can only be achieved by burning a fuel with a low sulphur content such as natural gas. The North Thames Gas Board has calculated that replacement of coke boilers by gas fired boilers in five hundred schools during 1970 will lead to a reduction of 99 per cent in the amount of sulphur dioxide released to the atmosphere by these premises.

There is still considerable scope for reduction of industrial contributions to air pollution. Large numbers of factory and power station smoke-stacks still belching out wastes of many different hues lend support to criticisms that have been made about the apparent complacency of the Alkali Inspectorate. The attitude of the Chief Inspector to the fact that cement emissions present 'no obvious danger to health' shows an appalling lack of understanding of the generally undesirable nature of such fallout on washing, paint-work and vehicles. Perpetuation of this type of situation means the continuation of large amounts of air pollution solely because there is no proven medical danger, completely irrespective of other undesirable features which might exist

The Alkali Inspectorate is responsible for checking over three thousand processes in Great Britain and operates on the basis of co-operation with industry to ensure that the 'best practicable means' of control are used. All too often the Chief Inspector's annual report gives the impression that the best practicable means is a higher chimney stack. In the three years 1967–9 only four legal actions were taken against companies. At the same time, of fourteen viscose works in the country only eight treat their emissions. The remaining six give no prior treatment before dispersing from high chimneys – surely this does not represent the 'best practicable means'? In a similar way, if a small number of oil refineries can remove 145 tons of sulphur from crude oil every day, it cannot be the 'best practicable means' for the remaining refineries to release 320 tons of sulphur per day (as SO_2) to the atmosphere. For all the claim that Great Britain is ahead of other countries in air pollution control, it is still quite common to see heavy plumes of emission from brickworks, power stations and cement works.

A third region in which action can be taken to reduce pollution is with vehicle exhausts. Currently, emissions of unburnt hydrocarbons, carbon monoxide and lead fuel additives go uncontrolled in most countries throughout the world. Legislation is several years behind the technical developments which have now advanced to the stage whereby the greater part of these emissions can be stopped. It is already apparent that several nations are now in the process of following the pioneering spirit of California and the U.S. Federal Government in applying controls to motor vehicles.

It is worth considering in some detail the action that has already been taken, and is in the process of being taken, to reduce air pollution. In Great Britain action followed rapidly on the heels of the 1956 Clean Air Act that enabled local authorities to establish smoke control zones. Within a period of ten years all but twenty-three local authorities in the black areas designated by the Beaver Committee had commenced the introduction of such controls. Cities in which large areas of smoke control now exist have reported substantial improvements in atmospheric conditions. These improvements have been particularly noticeable in London, where the smog disasters of 1952, and the lesser one of 1956, helped considerably in overcoming the lethargy of centuries. Now more than 60 per cent of London is controlled and the objective is complete smoke control by 1978. As a result of the action already taken smoke has decreased by 75 per cent and sulphur dioxide by 40 per cent. An additional gain is a doubling in the number of species of birds to be found in the London area. It has also been claimed that the resulting increase in sunlight has enabled Londoners to grow plants which could not have survived in the atmosphere of ten years ago.

Similar improvements have been reported in Sheffield, which is now claiming to be the cleanest industrial town in Europe and expects to attain full smoke control by 1972. At present 70 per cent of the city is controlled and further improvements will deal mainly with the industry within the city boundary. Already in that area smoke has decreased by over 60 per cent and sulphur dioxide by 25 per cent. The benefits this has brought to the residents include better visibility, cleaner buildings and relief to sufferers of bronchitis.

Because introduction of control of this type costs money – up to £50 to convert each household grate (totalling over £7 million already in London) – the central Government guarantees to cover the major part of the cost. A householder in a newly created control zone is only expected to pay 30 per cent towards the total cost of any necessary conversion. The local authority is also responsible for 30 per cent and the central Government pays the remaining 40 per cent. The combination of the 1956 Act and this form of financial support has already gone a long way to encourage local authorities to create smoke-control zones. Also slum clearance of thousands of old coal-burning houses, and their replacement with modern centrally heated flats, has contributed greatly to the improvements in many of the traditionally grimy northern cities like Manchester and Salford.

Unfortunately it is not all a success story. The 1956 Clean Air Act left the responsibility for creating smoke-control zones to the local authorities. This automatically left a loophole for a few areas in which local councillors were not prepared to take the necessary steps to reduce air pollution. By 1970 there were still twenty-three authorities in Great Britain that had refused repeated ministerial requests to create control zones. It was partially because of this recalcitrant attitude that the later 1968 Clean Air Act had written into it powers enabling the then Ministry of Housing to force local authorities to introduce smoke control.

In March 1970 *The Times* ran a very interesting article on the areas which had taken no effective control steps. It found that most of these authorities were in the north and a number were mining areas in which the mine workers were given a free allocation of coal to burn at home. Any imposition of smoke control will automatically necessitate changes in this tradition of nice coal fires. It obviously does not meet with the approval of a number of local councils which include a proportion of mineworkers amongst their members. Even during summer months it is a common sight to see black smoke coming from the chimneys of the homes of the three hundred thousand miners receiving free coal.

In addition to those areas that refuse to change, the cleaner air programme has been delayed because some areas that wish to change have been unable to guarantee the supply of the necessary smokeless fuel for the controlled zones. This meant postponing new smoke-control zones at least until April 1971. The shortage of fuel has also resulted in a number of local authorities asking for permission to suspend control temporarily in over one hundred of the zones previously created. As a result of a substantial miscalculation of fuel demands during the 1969–70 winter period, smoke control not only stood still but actually took a step backwards during European Conservation Year, and the effects still existed in the winter of 1970–1. The teeth of the 1968 Clean Air Act will now need to bite even harder to recover the lost ground.

In Great Britain air pollution from bonfires and household incineration has not yet been restricted or controlled by legislation. Neither the 1956 nor the 1968 Clean Air Acts deal with the burning of garden rubbish, so that any legal action depends upon a bonfire causing a nuisance in the immediate locality. In California, the city of Los Angeles has prohibited the burning of domestic waste for more than ten years. Such measures, even in small areas, can only be successful if full facilities are available for other methods of disposal. To deal with this problem it will be necessary to increase the number of municipal dumps and to expand the scope of local refuse collections.

With domestic smoke control at least well in hand, and with apparently sufficient legislation existing to ensure that it continues, considerable attention is being directed towards industrial pollution to see what additional steps can be taken to reduce this. Three approaches can be made to the elimination of industrial pollution. These are: stopping the polluting process completely and permanently; improving or changing the process to prevent the annoyance in the first place; removing any pollutants before they are released into the atmosphere.

The first method is only of limited application in such cases as when combustion for heating purposes can be reasonably and economically replaced by electrical mains heating. It also applies in the case of national power stations when alternative methods of generating by nuclear energy (which has its own disadvantages, as explained in Chapter 9), or by tidal flow are available.

Changing or improving the industrial process is a more common procedure, particularly in the case of fuel combustion processes. As combustion involves the use of large quantities of oxygen from the air it is well known that inadequate supplies of air will lead to the incomplete burning of the fuel and the release of carbon monoxide gas and large quantities of soot. This can often be overcome by simply improving the oxygen supply. An alternative approach is to change to a fuel which produces a smaller proportion of pollutants, and for this purpose gas is preferable by far to practically all forms of solid fuel and most forms of oil heating. There is no doubt that changing to gas firing can lead to substantial reductions in quantities of sulphur dioxide released to the atmosphere and the almost total elimination of unburnt particles. Oil tends to create a slightly greater amount of sulphur dioxide than does gas, as it is not possible to remove totally the natural sulphur content during oil refining.

The diverse nature of industrial pollution makes it almost certain that a great deal could be eliminated by simply modernising particular industrial processes and installing up-to-date equipment. One example in which improved industrial processes have reduced pollution has been with blast furnaces in which better methods of loading the charge at the top of the furnace have reduced the number of releases of furnace gases to the atmosphere. Most of the fumes formed in blast furnaces are in any case cleaned and recirculated to fire the furnace during the course of operation. Improvements in steel-making have also led to reductions in pollution. The traditional Bessemer and Open Hearth furnaces are responsible for the release of between 10 and 20 lb of dust for every ton of steel manufactured, whereas the more modern basic oxygen furnaces fitted with ancillary equipment are claimed to remove at least 95 per cent of the dust formed in the process.

The chemical industry in most countries has become very pollution-conscious during the past few years, and it is a well-accepted maxim that the existence of offensive smells or particles around a chemical factory indicates a poor chemical process that must be losing the company some of its profit. For this reason alone the chemical industry is normally only too anxious to take steps to avoid this form of wastage. One well-known example of an improved process making a significant saving has been in the manufacture of sulphuric acid, which was formerly only 94 per cent efficient, the remaining 6 per cent of the sulphur being released as sulphur dioxide to the atmosphere. More modern equipment is able to bring about a reaction efficiency of 99.5 per cent, giving a reduction of over 90 per cent in the amount of sulphur dioxide released, and even this may be improved by the fitting of ancillary equipment to absorb any escaping fumes. It has been estimated that the chemical industry in Britain alone was spending an average of £2 million per annum between 1958 and 1968 in installing new pollution-control equipment, and the running costs had attained nearly £7 million at the end of that period.

Paper- and pulp-manufacturing have long been regarded as major sources of pollution, mainly because of the numerous processes involved in the manufacture, and also because of the unpleasant character of some of the by-products that result. Particularly obnoxious and penetrating are the sulphur-containing compounds, hydrogen sulphide and mercaptans (organic sulphur compounds). The odour of mercaptans can be detected in concentrations as low as one part in one thousand million, but is not believed to be toxic in any way at this level. Quantities of sulphur lost to both the atmosphere and to sewers during the various paper-making steps can often be substantial, amounting to 20 or 30 lb of sulphur for every ton of pulp produced. This is an expensive loss for the industry itself and control processes have been developed and introduced. These include oxidation of the sulphurous gases followed by subsequent chemical absorption and condensation of pollutants during waste-heat recovery from hot effluent gases. Problems of pollution in this particular industry, however, are far from solved, and may eventually necessitate a completely new pulping process in which sulphur plays no part, as this is the major culprit.

These are just a few of the cases in which pollution has been arrested by improved processes. It is not possible for every manufacturer to purchase and install the most modern equipment overnight, either from a simple economic point of view or from an equipment supply point of view. In many cases small companies could be put out of business if legislation forced them to install modern processes at a moment's notice. Whilst this might lead to an immediate improvement in the atmosphere, it would have undesirable side-effects including increased unemployment. Because short-term improvements of this nature cannot be brought about, the third method of dealing with industrial pollution – removing the pollutants from the effluent gases before they are released into the atmosphere – must continue to figure prominently.

Control of this type has led to very sophisticated equipment being developed for the collection of particulate matter and waste gases. Where these methods are fully applied and effective the level of pollution can be reduced by at least 90 per cent compared with the uncontrolled process.

The simplest particle-removing method is the use of a gravity settling chamber. This is basically a large tank in which the issuing gas velocity is substantially slowed down. A series of baffles are fitted into the tank so that there is no straight-through route for the gases. During the course of its passage through the tank solid particles and droplets settle out due to impact and normal gravity action. This type of equipment is only adequate for preliminary treatment of the emission in which it is desirable to remove large particles. Removal of particles smaller than 50 microns requires better methods.

Many devices for removing solid particles work on the centrifuge principle such as a spin dryer in a washing machine, in which the spinning action of the tub forces the water out through holes in the side. Cyclone collectors make use of this principle by driving the gases under pressure in the form of a vortex through a cone-shaped tube. As a result of this treatment any particles are thrown outwards and collected below the orifice while the gas proceeds for further treatment before being released. In some cases it is the practice to have two or three cyclones in series feeding one to another in order to achieve a high level of particle removal. More commonly they are used in conjunction with other equipment, as most industrial emissions are mixtures necessitating a variety of techniques to remove the different components.

One very common approach to source-control of air pollution has been the use of liquid scrubbers, in which the effluent gases are passed through a liquid spray which removes the contaminants from the gas flow. A large number of scrubbing devices have been successfully developed and some are capable of removing particles as small as 1 micrometre in diameter. Most of these operate on the counter-current principle in which the gas flow is in the opposite direction to the liquid or solvent spray. The simplest form of this is a shower device spraying into the head of a long tower into which the gas is fed at the base. The purified emission is removed at the top of the tower. Such a method is only moderately efficient, and various modifications have been devised to improve upon this. By packing the column with inert substances such as glass balls, metal spirals or porcelain rings, it is possible to increase the effective surface area of the liquid and remove a larger proportion of the smaller particles which would pass through a more simple arrangement. The best forms of this method of purification spray the liquid at right-angles across the direction of flow of the gas and can operate at very high flow capacities.

The spray liquid employed in scrubbers is frequently water when the particles to be removed are reasonably water-soluble. In some cases, however, dilute solutions of chemical reagents are employed to react with the gaseous components which might not dissolve sufficiently in water alone. On occasions the spray liquid is a very fine suspension of an active material such as carbon which is capable of adsorbing substantial volumes of most impurities. In all scrubbing procedures the efficiency of the method decreases rapidly if the gas flow is too rapid, and careful control of both gas and spray must be maintained for the best results to be achieved.

It should be pointed out that the very removal of pollutants by spray techniques automatically creates additional problems, as in most cases the contaminated spray liquid is unsuitable for discharge into rivers or sewage. In some cases the scrubbing liquid can be recirculated through the scrubber; alternatively any substances it has extracted may be precipitated before the waste liquid is discharged, to avoid creating a water pollution problem as a result of having solved an air pollution one.

Removal of pollutants by filtration is the traditional method of dealing with particles and aerosols, and has been greatly improved by developments of new fibres and fabrics suitable for long-term operation. Particles with diameters as small as one-tenth of a micrometre can be removed by the best filters, although it is more common to think in general terms of removing particles of about 1 micrometre diameter. The most important part of a filter unit is the filter bag, which must not tear under the flow pressures of the gas and must be resistant to chemical attack from the impurities which it is intended to remove. Because of these criteria, synthetic fibres now play an important role in this field. Fibres such as nylon, orlon and glass meet many of the requirements because of their resistance to chemical attack, although the success of removing particles is really dependant upon the build-up of a layer of particles on the material rather than on the closeness of weave of the fabric.

Many filters operate by forcing the contaminated gas flow into one end of a long tube made from the filter material. Both ends of the tube are open, one end taking the inlet gases and the other end fixed over a waste hopper. The flow pressure forces the gas through the walls of the fibre tube and the solid particles are retained inside the tube. As the layer of removed particles builds up it can be dislodged to fall into the hopper below from which it can be periodically removed. This cleaning is necessary, as a thick layer will result in a considerable reduction in the flow rate and a corresponding build-up of pressure in the purification equipment.

Apart from the use of filter bags there is also air purification equipment that utilises fibre beds or mats. These are of special application for air-conditioning and ventilation purposes where the level of impurities to be removed is relatively low. They are also used for preliminary treatment of polluted atmospheres to prevent overloading more expensive equipment used for smaller particles.

Closely related to filter units, but operating on a different principle, are gas-solid adsorbers. These are based upon the fact that certain solids have a great ability to adsorb, but not necessarily to react chemically, with a wide range of gases and vapours. The best known of these substances is activated carbon, prepared from bituminous coal or wood, and which is used extensively in filter beds to remove quantities of organic solvents from the atmosphere. In industrial applications carbon or metallic oxide adsorbers have to be quite large to cope with substantial air flow rates which are often of the order of 1,000 cubic feet every minute. The filters may be in the form of cylinders or folded sheets formed by depositing the adsorbent upon a flexible support. In these cases the filter bed is a few inches thick, but flat bed adsorbers about 2 feet thick have been used to remove large quantities of organic solvents from effluents.

A domestic application of this form of air-purifier has been in household deodorisers, in which a circulating fan is used to force the air in a room through a block of activated carbon. Cigarette smoke, food smells and body odours are efficiently removed in this manner. Industrial applications are of much greater importance, particularly for removing penetrating odours detectable at concentrations below one part per million. Pharmaceutical processes, food preparation and the tanning industry all give scope for this method of purification to avoid releasing undesirable smells into the atmosphere.

Another method for removing solid and liquid particles from a polluted air flow is by means of an electrostatic precipitator. As long ago as 600 B.C. the Greeks were familiar with the electrostatic attraction that could be imparted to amber rods by rubbing with cloths. The English court physician William Gilbert mentioned the attraction which electrically charged objects have for particles of smoke in A.D. 1600, but research in this subject did not really start until the end of the eighteenth century. The first attempt to apply electrostatic precipitation was carried out at Bagillt, North Wales, in 1885, but was not successful. It was not until the early part of this century that really practical devices were invented by Frederick G. Cottrell.

The technique is based upon the concept that small moving particles can be induced to pick up an electrical charge and hence be attracted to an electrode possessing an opposite charge. An electrostatic precipitator consists of a series of wires each suspended along the centre of a vertical tube. A permanent electrostatic field is maintained between each wire and tube and attracts particles in the gas stream as they pass along to the walls of the tubes. The particles which aggregate on the tubes are shaken loose by periodic mechanical rapping and are collected in a hopper at the base of the unit. Particles less than 1 micrometre in diameter can be removed with an efficiency greater than 99 per cent. As demands for cleaner emissions increase, more and more electrostatic precipitators are being installed to deal with industrial problems. They already play a major part in removing dust from cement manufacturing and particles from refuse disposal plant incinerators, and will certainly continue to be one of the most important methods available for cleaning the atmosphere.

It is uncommon for just one form of purifier to be used in isolation to treat gaseous emissions which are intended to be released to the atmosphere. A combination of two or three different units fixed in series is frequently necessary to clean the air sufficiently to satisfy even moderate standards. A typical arrangement would have a gravity settling tank as the first stage, to remove coarse particles; either a liquid scrubber or cyclone collector as the second stage for finer particles; and electrostatic precipitators to remove as much of the residual particulate matter as possible.

With the large amount of equipment now available for dealing with industrial pollution there is little excuse for any power station or manufacturing process continuing to emit large quantities of dust and muck, whether or not these are harmful to health. There is now a moral obligation for such industries to take greater steps to deal with their pollution. Whilst expense prevents rapid short-term solutions from being possible there is no reason why a particular manufacturer cannot work on the basis of a five-year programme progressively to reduce and eventually eliminate the release of pollutants. This can only be fully achieved by greater pressure from the Alkali Inspectorate, or by further legislation directed more particularly against industrial pollution.

It is quite obvious to most people that the motor car is a major source of air pollution and it is difficult to understand how this has virtually escaped control at a time when householders and industry are being made to take steps to stop contaminating the atmosphere. A considerable difference now exists between the action already taken to control vehicular emissions in the U.S.A. as compared with Great Britain. First steps to stop vehicle emissions anywhere in the world were taken in the State of California in 1960, when a law was introduced requiring control devices on new motor vehicles from 1961. This independent state action was followed by a 1965 Federal amendment to the U.S. Clean Air Act of 1963 authorising the Secretary of Health, Education and Welfare to establish emission limits for all motor vehicles on a national basis. In Great Britain the only control at present existing is that against the emission of black smoke by diesel exhausts, imposed by the Motor Vehicles (Construction and Use) Regulations, 1966.

It has already been pointed out that motor vehicles emit a number of pollutants which under certain climatic conditions will produce photochemical smog. These emissions fall into three groups requiring different techniques to overcome them. Firstly, the main source is obviously exhaust gases. Secondly there are crankcase emissions arising from the escape of gases from around the piston rings in the engine. Thirdly there are emissions due to evaporation from the fuel tank and the carburettor.

Pollution from diesel engines is relatively small, as they only produce about one-tenth of the carbon monoxide emitted from a comparable petrol engine. The design of the engine and fuel supply also lead to less fuel evaporation into the atmosphere. In all diesel engines the main problem is that of controlling smoke emissions. This can be partially achieved by adding a minute amount of barium compound to the diesel fuel. In all cases improvement can be achieved by correct, regular maintenance of the fuel supply system and also by stopping the use of the excess fuel device. Vehicles observed

Much river pollution is caused by major towns discharging untreated and partially treated sewage into them. Fortunately this sewage flow, photographed in the centre of Newcastle upon Tyne, should soon be a thing of the past as an enormous sewage disposal scheme costing £35 million is being planned for Tyneside.

Aeration of sewage effluent, such as this shown at the Bellozanne treatment works in the Isle of Jersey, is an essential feature in the complete sewage treatment process. The failure to break down synthetic detergent is one result of inadequacy of much sewage treatment. This is particularly obvious in polluted rivers which can frequently be covered with mountains of froth.

The visual deterioration of the environment is clearly demonstrated by these contrasting views of the famous Manhattan skyline. On a clear day there is no difficulty in seeing across the Hudson River, but on days of severe atmospheric pollution visibility is restricted to a few blocks.

In Britain a great amount of the research into air pollution is carried out at the Government's Warren Spring Laboratory at Stevenage.

The combination of air pollution and industrial ugliness, as seen at Ebbw Vale, Wales, is an all-too-common feature in the developed nations of the world.

to be emitting black smoke are usually underpowered or overladen, and one obvious answer to this problem is tighter legislation covering power-to-weight ratios for fully laden lorries.

It is a mistake to assume that these sources of pollution are either too small or too difficult to be adequately dealt with. Individually one vehicle contributes only minutely to the total atmosphere. But it is not one car that is the cause of the problem. In Los Angeles alone there are nearly four million motor vehicles all emitting fumes, whilst in Great Britain, which is only about the size of California, there are about 15 million registered motor vehicles adding to the pollution. Some estimates suggest this figure will double during the next ten years. As there is no current legislation forcing British car manufacturers to take steps to control emission on vehicles for the home market, Great Britain is well behind the U.S.A. in this respect. As it is, British manufacturers are having to fit modifications to make their cars saleable in the U.S.A., but still continue to manufacture mobile pollution machines for this country and the rest of the world.

It is, of course, fair to say that the cost of vehicle pollution control means an increase in the price of a new car. It would be economically impossible for a single manufacturer to fit all its vehicles with the necessary modifications as they would be priced out of the market. In the light of the absence of government legislation manufacturers should make pollution control units as optional extras on new cars while arranging between themselves to promote and encourage overall improvements in design to reduce pollution. Within another five years the number of countries restricting vehicle emissions is going to increase substantially, and it is well known that the manufacturers are already carrying out further research in preparation for the days when it will be an offence to sell an uncontrolled vehicle in any country.

Methods for control are already well established in the U.S.A., and are beginning to contribute to a reduction in atmospheric contamination. Control of crankcase emissions leads to a reduction of about 20 per cent in the quantity of hydrocarbons released by a motor car. This can be achieved by running a tube from below the piston back to the carburettor so that any unburnt hydrocarbons from this source are recycled instead of escaping. Control of this type started in California with the 1961 models and was extended to models sold nationally throughout the U.S.A. in 1963. It is certainly one of the easier and cheaper pollution-control steps that can be made on motor vehicles. Compulsory recycling of crankcase emissions will apply to new vehicles in Great Britain from October 1971.

Exhaust pipes are intended to carry away the products of combustion to the atmosphere. If the combustion is 100 per cent efficient these will simply be carbon dioxide, nitrogen and water; none of which is considered to be a pollutant. No car engine attains this utopian standard, so the exhaust gases always carry with them the products of incomplete combustion, and it is these that must be removed.

Any device should therefore be designed either to absorb these products or to complete the combustion process. Motor car manufacturers have not found it easy to meet the requirements being imposed in the U.S.A., and the levels are due to be tightened further during the next few years. Improved engine design has enabled some manufacturers to overcome the first hurdle, whilst others have injected extra air into the very hot part of the exhaust where it is attached to the engine at the manifold, enabling the combustion to proceed further. These methods are unlikely to be adequate when the standards are made more restrictive.

Modifications now being developed include manifold reactors to improve upon the present system described above, and after-burners fitted to the tail-end of the exhaust pipe. The latter device allows an additional flow of air to mix with the exhaust gases, followed by ignition to spark off the extra combustion necessary to oxidise the carbon monoxide and hydrocarbons. An alternative approach to igniting the mixture in the special after-burner attachment has been to pass the air and waste gases mixture over a catalyst which enables the oxidation to occur at a lower reaction temperature. Unfortunately most catalysts have only a limited life, and the converters developed so far need replacing every 6,000 to 10,000 miles. This has been because the catalyst is poisoned, owing to the presence of lead from the anti-knock additive in the exhaust gases. To overcome this problem fully it is necessary to stop putting lead compounds into the petrol at the beginning. This is in any case desirable, as lead is itself a poison and has been shown to be responsible for stunting the growth of roadside plants.

A number of lead-free petrols are already available, but are generally only suitable for older low-compression car engines. Production of lead-free petrol suitable for modern high-compression engines would cost the motorist between 1p and 2p per gallon extra. It has been estimated that the eventual reformulation of petrol to meet the demands of improved engine design and the elimination of lead compounds is likely to cost over £100 million throughout the world. This financial burden will for the greater part undoubtedly be transferred to the motorist, and the speed with which it comes will depend upon how rapidly the nations legislate on the matter of vehicle pollution. Despite these difficulties, the U.S. Government has passed legislation aimed at reducing exhaust pollution by 90 per cent by 1975.

One group of exhaust gases requiring special attention are the nitrogen oxides formed from the reaction of atmospheric nitrogen with oxygen in the car engine. Although the level of these oxides is very low they play such an important role in the photochemical smog reaction that it is imperative that every reasonable step is taken to prevent their formation or to ensure their removal. The main method of control at present favoured again involves recycling part of the exhaust gases back to the engine. This leads to a reduction in the combustion temperature of the engine with a corresponding decrease in the amount of nitrogen oxides formed.

It will be appreciated that many of the steps necessary to stop pollution from motor vehicles involve improved engine design. Work in this field has been carried out for a number of years, but worth-while results will probably only be achieved by substantial investment being made in engine development in the years ahead. This will only really come about when public opinion is strong enough to bring its influence to bear on both manufacturers and legislators – it cannot take place until there is an overall improvement in our personal attitudes to pollution and waste disposal.

Of course, research is also being carried out on alternative methods of propulsion for motor vehicles. Considerable impetus was added to this by the 'clean air' race for cars from Cambridge, Massachusetts, to Pasadena, California, in August 1970. The competition attracted fifty entries, ranging from electric cars to steam engines. Many of these were up-dated versions of old concepts. Electric cars were tried in the nineteenth century and one version, *La Jamais Contente,* attained a speed of 65·7 m.p.h. in 1899. In Great Britain electric propulsion has been used in delivery vehicles such as milk floats, while an American inventor has developed a pollution-free engine which runs on continuously recirculated gas obtained by heating a refrigerant called Freon. It has been claimed that the engine may be used in vehicles by 1972.

Some critics have complained that the anti-pollution campaign is based heavily upon emotion and only partially upon reason and scientific appraisal. People who see their washing rotting away from acidic vapours and their children covered with dirt smuts have a right to get emotional. Households compelled to buy expensive smokeless fuel cannot appreciate any so-called scientific reasoning when industrial smoke-stacks continue to belch out their poisons. Further successes in overcoming air pollution will depend upon all sections of the community working more closely together for the benefit of everyone.

7 Clearing the dustbins

Once a week an anonymous individual walks up the path of the majority of houses in this country and proceeds to empty the dustbin and remove the week's accumulation of waste paper, empty bottles, food residues and other domestic refuse. The 'refuse disposal operator' is an unskilled workman, relatively poorly paid, and carrying out a menial task. But failure to do this job results in piles of decaying rubbish being dumped indiscriminately in the streets; as occurred during the dustmen's strike of 1969, and the 'dirty jobs' strike of 1970. It is such situations that make people conscious of the vast quantities of material that they discard each day as refuse, much of it arising from contemporary methods of packaging.

In one year the average person produces between 600 and 700 lb of household waste, which for a family of four would create a pile between 6 and 10 feet high if all dumped together. Because refuse collections are, for the greater part, well organised, the full magnitude of the task imposed upon local authorities is rarely appreciated by the general public. Not only must the collections from domestic, distributive and industrial sources be maintained at frequent intervals, but there must also be adequate methods available to salvage and dispose of the wide range of materials obtained as a heterogeneous mixture of paper, tins, glass, rags and plastics.

As standards of living rise so the quantity of refuse increases and the proportions of different materials change. For example, the proportion of dust and cinders which used to comprise over 50 per cent of the domestic refuse only ten years ago is estimated to have decreased to 30 per cent, mainly due to the advent of smokeless fuels and central heating. At the same time the proportion of paper and cardboard has risen during the same period from 16 per cent in 1960 to an estimated level of 33 per cent in 1970. In addition to this there has been a significant increase in the amount of plastic materials being discarded, which gives rise to further disposal problems. The present indication is that quantities of cinders and dust will continue to decrease whilst the amounts of packaging materials will increase. Food wastes appear to maintain a fairly constant level, as do waste cloths and rags. In recent years many local authorities have taken responsibility for disposing of large items such as old furniture, mattresses and carpets, and these also have necessitated special treatment.

Obviously the adequate disposal of solid waste is essential for the health and welfare of the nation. Evidence of indiscriminate dumping of rubbish by many people is only too obvious from the numbers of old gas stoves, oil cans, mattresses and vehicles that lie slowly decomposing in our woods and hedgerows. Roadside graveyards for dead motor cars exist all over the country, often systematically dissected in public view by human vultures who have no concern either for the mess they leave behind for others to clear up, or for the dangerous playing environment they create for their own children.

But the sad fact is that the average individual is, in his own way, just as inconsiderate. There needs to be a considerable change in public attitude to prevent such places as picnic sites, laybys, and parks from disappearing under a mounting pile of waste paper, beer cans and broken bottles. The ubiquitous motor car has enabled human litter louts to penetrate to even the most isolated parts of the globe where they continue to make their presence felt after departing by leaving behind the disgusting residues of so-called modern living. The result is considerable expenditure in time and money for local authorities to remove rubbish which should not have been left in the first place; even they cannot always cope and the countryside is littered with more or less permanent dumps in out-of-the-way places. A little bit of thought and consideration is all that is needed for this to be avoided.

The Keep Britain Tidy Group, which has been campaigning against the thoughtless and indiscriminate dropping of litter for more than sixteen years, has drawn attention to the considerable expense involved in cleaning up after the litter louts have been around. For example, it took 170 men with 8 lorries a complete day to clear the rubbish from Wembley Stadium after the 1968 World Cup Final. In the town of West Hartlepool it has been estimated that the cost of keeping the streets clean could be reduced from £40,000 to £25,000 if the public would use the street litter bins provided.

Municipal waste disposal, when correctly developed and organised on modern lines, is well capable of handling all of the refuse created by contemporary living. There is no need for anyone to discard rubbish in other than an authorised place and manner. Disposal methods for domestic waste include sink disposal units, land reclamation, separation, incineration and composting. In some cases a combination of these methods is used, and is at present costing Great Britain £60 million per annum.

Sink disposal units are common in the U.S.A., being fitted in 15 per cent of households, but are still relatively new in the United Kingdom. The unit is an electrically operated masticator which fits below the sink outlet and shreds up food residues as they are fed into it. The particles are flushed into the waste pipe to join the other water-borne wastes from the household for treatment at the sewage works. About 10 per cent of waste, mainly food residues, can be treated in this manner. Its main advantage is that it avoids having decomposing waste in a pedal bin or dustbin for several days between collections. The disadvantage is mainly one of expense, as purchase and installation of a unit may cost as much as £100. Sink disposal units are banned in certain towns in the U.S.A. because large numbers in any particular area can lead to overloading in the local sewage works if it is only of limited capacity.

Another approach to home destruction of refuse has been the development of small gas-fired incinerators. These are now banned in enlightened areas such as California because of the large contribution they make to air pollution in most cases. Some types are banned in smoke-control zones in the United Kingdom for the same reason. They are effective in reducing the large bulk of domestic refuse, but, like the sink disposal units, still leave a large proportion of the rubbish to be disposed of in other ways.

As much of the bulk of modern waste arises from the increasing quantity of resilient plastic packagings it has been suggested that the volume could be reduced by the development of a household shredder or disintegrator. This would certainly mean that dustbins would take longer to fill and could at the same time reduce the cost of municipal waste collections by making only one call every two weeks necessary.

The objectives of local authority waste disposal systems are: to reduce the bulk volume of the waste it receives; if possible to cover some of the operating costs by selling recovered or useful materials; to ensure that any remainder is made as innocuous as possible; and to dispose of the residue in a suitable manner.

The simplest method of dealing with this has been to hand-sort the waste to remove useful paper, ferrous and non-ferrous metals, glass and rags and, after sieving the remainder, to remove ashes and dust for brick manufacturing, and to use the residue for land filling. Simple tipping of waste into cuttings or abandoned gravel pits is undesirable, as it results in a very slow decomposition accompanied by the production of unpleasant smells. The alternative method, which is used extensively for land reclamation, is to sandwich the screened and sorted refuse between alternate layers of earth. A 2- or 3-foot layer of earth between every 6- to 10-foot layer of refuse means that not only can the land level be substantially raised, but the decomposition takes place without causing offence to local residents. The limitations to this form of disposal are the cost of transporting refuse to sites more than a few miles away, and the increased difficulty of finding suitable dumping and reclamation sites. Considerable care has to be taken in selecting land-fill areas because of the possible dangers from leaching and rain-water run-off that can pollute rivers. This is of particular importance with the dumping of toxic industrial wastes. It is, in fact, becoming increasingly necessary for solid wastes to be dealt with by methods other than land reclamation. Any plant constructed for the purposes of separation, incineration and/or composting must also comply with the regulations against air and water pollution, as it is not sufficient for the solution of one waste problem to create another.

The process for separation and incineration starts with the tipping of the waste from the lorry into a hopper which feeds the material on to an inclined conveyor belt. This elevates the refuse to deposit it on to a vibrating or rotating screen to free it from dust and ashes. From the screening the sieved refuse is transferred to another conveyor belt for hand-sorting to remove saleable items of refuse. Many items which were well worth recovering years ago can no longer command a price to justify the expense of hand-sorting, so that often only non-combustible materials such as glass are now hand-picked. The conveyor belt next passes under an overhead magnetic separator which attracts any ferrous metals to it. This is mainly in the form of empty tin cans which can be easily compressed into dense blocks measuring one cubic foot for sale as scrap iron. In some instances unbaled empty cans can be treated to recover the thin tin coating before using the scrap iron, but this involves the expense of transporting bulky empty tin cans to the scrap dealer, and is only an economic proposition if there is a local de-tinning works. In modern refuse treatment plants the waste is automatically fed into a moving grate furnace in which the incineration is carried out at temperatures up to 1,000° C.

Combustion of 3 tons of refuse produces about the same amount of heat as 1 ton of coal, and this heat can be used to drive steam turbines for generating electricity. Many modern refuse plants not only provide their own requirements for electricity and heating but also sell any excess electricity generated to the National Grid. It has been calculated that a new incinerator in Montreal will consume 1,200 tons of refuse each day and produce 400,000 lb of steam per hour. Because of the wide range of materials in refuse, combustion automatically leads to the production of most common forms of air pollutants. Modern plants are equipped with cyclones and electrostatic precipitators to reduce this pollution to a minimum. Many of the complaints levied against municipal refuse plants are because of ash and particles released, usually by older, out-dated equipment.

To overcome deficiencies of older plants, many of which are of limited capacity, local authorities are beginning to co-operate far more to construct large plants capable of taking refuse from a large area. Tunbridge Wells Borough Council, for example, has constructed a refuse disposal plant capable of dealing with three times the amount of rubbish produced within the Borough. This cost £200,000 and will probably serve a large number of the other local authorities in the same area.

Undoubtedly the largest and most expensive disposal plant in Great Britain now is the G.L.C. plant at Edmonton. This cost £10 million and started operation in August 1970. It is designed to take 1,300 tons of refuse every day (about one-eighth of the daily total of London's waste, delivered in over 600 lorry loads). Generation of electricity in this plant is expected to raise nearly half a million pounds a year to be offset against the running costs. In line with contemporary views on the environment this plant is not only to be screened by trees, but the area around is being landscaped to avoid creating a blot on the edge of the Lee Valley Regional Park. In addition to this the plant is fitted with wet dust extractors and electrostatic precipitators to avoid air pollution. Water used for generating is mainly recirculated town water and the cooling water is purified sewage effluent. This combination prevents any great demand being made on the local water supply and limits the possibilities of creating any water pollution. Construction of other large incinerators of this type should eventually obviate the need to continue the use of old, inefficient plants.

Despite this, the Greater London Council will still continue to use some of its waste for land reclamation where this is suitable and convenient. Alternative treatment by composting refuse is still used extensively throughout the world. The justification for this type of process is usually dependent upon a continuous local demand for the compost formed. Natural decomposition of domestic rubbish can take up to several years, and occurs with any discarded pile of debris containing organic materials. The process of decomposition, brought about by the action of aerobic bacteria, can be speeded up and it is the purpose of modern composting plants to provide the optimum conditions for this process to take place.

Composting takes place after the rubbish has been screened, hand-sorted and the metals separated. It is passed through a pulverising machine which reduces material with a large surface area to below a standard maximum size ready for the fermentation process. At this stage the material for composting consists mainly of vegetable and animal wastes with a proportion of bits of paper and often plastics; a high organic content is essential for a good compost to be formed.

The decomposition process carried out by the bacteria requires an adequate supply of moisture and oxygen in a warm environment. These conditions are attained by either turning the refuse for several days in a rotating drum while air is blown continuously into it and water supplied as required. Alternatively the refuse is spread into large trays arranged in floors one above the other in a tower, so that refuse loaded at the top can be progressively transferred to lower trays during the time the fermentation is taking place. In all forms of fermentation a temperature between 60° and 70° C is attained. This is necessary to kill off various harmful micro-organisms such as *Salmonella*, which causes gastro-enteritis, and other bacteria responsible for diphtheria and typhoid fever which can exist up to about 55° C. The complete process takes about one week.

In some fermentation plants the organic level of the compost is raised by adding sewage sludge to the waste before the fermentation process. This was the original purpose of the composting plant constructed in the island of Jersey during 1959. It operated on the fermentation tray principle, using six floors. It was found that this gave rise to odours from the resulting compost and has since been discontinued. Production of compost without incorporating sludge has been proceeding satisfactorily for many years now, and nearly ten thousand tons of compost were supplied to Jersey farms in 1969. This particular fermentation plant was a prototype at the time it was constructed, and was the forerunner of a much larger version constructed in Bangkok, which is possibly the largest municipal composting plant in the world. It is claimed that the compost formed in this way is comparable with farmyard manure.

Because of the vast quantities of waste materials that have to be dealt with throughout the world, there is considerable research being carried out into new methods for dealing with refuse. Japan has as great a problem with rubbish as other industrial countries, but one company at least has proposed a different form of treatment. The idea, developed by the Tezuka Kosan Co., subjects domestic waste, along with scrap metal and paper to a uniform pressure of 3,000 tons. By this means stable blocks of compressed wastes weighing up to 5 tons are formed. It has been claimed that bacterial decomposition is impossible and the blocks can be used in numerous constructions. The process has already aroused international interest and could well become an important method for rapid disposal of waste.

Another technique under investigation that looks very promising has been developed by the United States Bureau of Mines. It has found that domestic waste, including sewage sludge, heated with water and carbon monoxide in a sealed container produced oil very similar in character to fuel oil. The oil produced has the added advantage that it possesses a very low sulphur content. It has been calculated that two barrels of oil can be obtained from one ton of refuse in a reaction carried out at 250° C and a pressure of 1,500 lb per square inch. This development is still in its infancy, and, like many other good ideas, will depend for its success on its financial viability compared with established methods of waste disposal. It has the advantage of creating a saleable end-product.

Dealing with waste is a very expensive operation. Running costs alone for refuse disposal are about £1 per head of population per annum, and are likely to increase further as more and more cities install modern equipment. Even a moderately sized plant can cost £1 million to construct, and only a small proportion of the running costs can be recovered from saleable items. A new refuse plant for the City of Nottingham, for example, will cost over £2 million: just to deal with the wastes from about 300,000 people. The cost of waste disposal is also likely to increase as a result of the changing nature of the refuse, which may necessitate adding to the number of steps in the screening and sorting processes.

A major problem is developing with the rapidly increasing quantities of plastic packaging now being used. These come in every conceivable form as plastic bags, expanded plastic liners, blow-moulded bottles and plastic utensils. By their very nature plastics are intended to be resistant to decay. If they are discarded in a haphazard manner they will remain relatively unchanged for years, unlike paper and even tin cans, which decompose quite rapidly.

In 1968 it was calculated that just over 1 per cent of all domestic refuse consisted of waste plastics, of which about three-quarters originated from packaging. There is every indication that the quantity of plastics used will continue to increase rapidly during the next few years. Currently Britain is well behind the U.S.A. and most European countries for the amount of plastic products used per head of population. It has been estimated that the annual increase will be between 10 and 12 per cent per annum, rising from 50 lb weight per head in 1970, to over 120 lb by 1980. Because of the non-destructible nature of plastic materials disposal problems will increase pro rata unless new techniques can be found.

Even the small quantities of plastics now arriving at the refuse plant for treatment are creating difficulties. This is particularly the case with composting plants in which the plastic pieces go right through to the composting stage and come out at the end of the fermentation unchanged. It means that the use of refuse compost can lead to an accumulation of plastics on the fields over a period of several years. In a similar manner large quantities of plastics in refuse used for land-filling can lead to a slowing down of the normal decomposition processes and the creation of only a loosely packed site. Another disadvantage of using plastics for land-filling is that lightweight products such as plastic bags and cups can blow away in windy conditions.

Currently plastics only account for about 8 per cent of the total of material used for packaging in the United Kingdom. This proportion will almost certainly rise very rapidly in the next few years displacing some of the traditional packaging materials such as paper and cardboard and resulting in a lower biologically degradable content of household waste. Whilst the current 1 per cent content of plastics in domestic waste is not a major problem, further attention will need to be paid to improved disposal methods as this level increases.

One approach to the problem has been to improve methods of pulverisation, thus ensuring substantial fragmentation of the plastic material and making it more suitable for land-filling and composting. In the long run, however, the most general method of waste disposal, particularly that containing a large proportion of plastic material, will be by incineration. This is the only method which leads to a sufficient reduction in both the volume and weight of the waste within a short period of time. Greatly improved incinerators have been developed in recent years which have made it much easier to carry out this process.

Most plastics consist of very large molecules in the form of chains of carbon atoms attached to other atoms, which are chiefly hydrogen or oxygen. Some polymers also contain nitrogen, sulphur and chlorine. The most commonly used plastic for packaging purposes is undoubtedly polythene. This has a slightly waxy feeling and is admirable for plastic bags and sheeting. It is made up entirely from chains of carbon atoms attached to hydrogen atoms, and on combustion only forms carbon dioxide and water. By comparison polyvinyl chloride, which is also a common packaging material, forms the corrosive acid gas hydrogen chloride on combustion.

The process of combustion of plastics is very complex and in some cases starts with melting and decomposition into simpler molecules at temperatures below 250° C. Some polymers develop a blistered appearance as the temperature rises due to the release of small quantities of gases. In the majority of cases combustion at temperatures up to 1,000° C leads to total conversion of plastic materials into gaseous substances that can be released into the atmosphere. Acidic gases formed during the combustion are known to accelerate the corrosion of incinerators under certain conditions. This can greatly increase the operating costs of this method of disposal and is an important factor that has to be borne in mind. Investigations have shown that the corrosive action can be overcome by operating at high temperatures and at the same time improving the flow of gases and air through the incinerator.

It is in any case necessary for all incinerators to be fitted with air pollution control equipment, which in this particular case should be capable of removing hydrogen chloride. The Alkali Inspectorate has recommended an effluent concentration of 450 p.p.m. of hydrogen chloride as being acceptable. This is far beyond anything that would be created by the present quantities of chlorinated plastics. However, many people would feel that even a level of 450 p.p.m. is too high when considered in conjunction with other 'small' quantities of pollutants released into the atmosphere. Because of this, research is being carried out to improve incineration procedures even further, and to reduce acidic emissions by causing them to react with other components of refuse during the actual incineration, thus preventing any release into the atmosphere.

A great deal of interest has been created in the possibility of forming plastics which can be readily degraded by microorganisms. This particular approach to the problem received national publicity from the first episode of the B.B.C. television series *Doomwatch*, in which the plastic-eating bacteria seemed to develop from an inactive state to being able to disintegrate a complete aeroplane in a matter of a few seconds. Plastics are used for many purposes for the very reason that, amongst other properties, they do not decompose during storage under the action of either bacteria or sunlight. Ravenous bacteria of the *Doomwatch* type, hungry for pieces of plastic, are not known, and this way out of the waste problem does not seem very promising. For some reason that is not at present entirely clear, plastics are not suitable for the nutrition of micro-organisms and cannot easily be broken down by them into more simple molecules.

Sunlight does have the ability to produce some discoloration of present plastics, but does not alter the mechanical properties to any great extent. If either bacteria or sunlight could achieve destructive results this would nullify the qualities that give plastics a long shelf-life as packages and containers. One British university is reported to have produced a plastic which crumbles to dust after a certain period of time. Full details of this have not yet been released. The discovery is based upon the idea of incorporating light-sensitive active chemical groups into the polymer structure. By this means the plastic is given a certain span of life after which a change in colour occurs before disintegration takes place. But the mind boggles at the very possibility of premature collapse of containers occurring on the shelves of a supermarket! The concept of chemical and biological degradability of plastics at present appears to create at least as many problems as it solves. For the time being, salvation from the tide of plastic bags appears to rest with improved incineration.

The disposal of waste is not restricted just to emptying dustbins and collecting garbage from shops. Many industrial concerns produce large quantities of solid wastes, sometimes of very toxic materials. Also, the ordinary household has garden refuse to dispose of in addition to old items of household furniture, motor vehicle tyres, discarded cookers, refrigerators and old motor cars. A great deal of the discarded litter disfiguring the countryside has arisen from the failure of local authorities to take the necessary steps to deal with the changing nature of national waste. The problem has undoubtedly been aggravated by some manufacturers deliberately pursuing the path of planned obsolescence, thus necessitating the regular replacement of broken or worn-out items. It is quite obvious that inadequate facilities exist for disposing of a great deal of the waste which is created.

One newspaper report is typical of the situation which can occur. To preserve the beauty of the Cotswolds, thirty special rangers spend their time removing the unsightly rubbish left by selfish tourists. In July 1970 this rubbish was still forming a gigantic pile in the back garden of a cottage at Stow-on-the-Wold because the North Cotswolds Rural District Council had refused the rangers permission to use the local dump. It is difficult to see how any pollution problems can be solved when this type of attitude prevails.

There is a need for local authorities to expand greatly the scope of the refuse collection and disposal services which they operate. For instance, very few councils currently have any arrangements for removing garden refuse. As a result householders are reduced to having to hide their grass-mowings surreptitiously in small quantities by wrapping them with their household waste in the dustbin. Alternatively they find some rural by-way where they can dump their leaves, grass-cuttings and prunings. Even those with garden incinerators are adding to the air pollution, and often annoying their neighbours into the bargain. As local authority disposal by either composting or incineration is greatly helped by a high organic content in the rubbish, garden waste should be an automatic item for inclusion. At the very least local authority sites for dumping this type of refuse should be provided.

The Litter Act, 1958, was the first attempt in Great Britain to deal with the haphazard dropping and discarding of rubbish in any public place. It enabled magistrates to impose a maximum penalty of £10 on anyone convicted under the Act. It would be fair to say that the Act has been almost totally ignored as convictions are very uncommon and most cities, towns and villages, and indeed much of the countryside are still disfigured by litter. Only about 4,000 prosecutions are made under this Act each year. Conviction results in little publicity, no public condemnation, and makes no impression on the true magnitude of the problem. The more serious aspects have now been covered in the Civic Amenities Act, 1967, but it appears that the only way to prevent litter-dropping will be by implementing on-the-spot fines, as is common in some European countries, and by increasing personal concern about litter. Under the provisions of the 1967 Act local authorities have a duty to establish sites in which householders can deposit refuse free of charge. Unfortunately many local authorities do not yet appear to have taken the necessary steps to provide such places. The finding of suitable areas in large towns and cities is certainly not easy, as wherever is chosen is likely to be opposed by local residents. This is, however, a very necessary prerequisite to preventing further spoilage of our diminishing countryside.

Similar problems arise over the disposal of large items such as furniture and carpets. It is only too easy for local councils to object to removing such items on the grounds that they do not constitute household waste. This has often placed an obligation on the local furniture stores to guarantee removal of old items of furniture in order to be able to sell the new. In any case the item no longer required has to be broken up eventually and disposed of by someone somehow. It is far better that this should be done in an approved authorised manner than that it should be permitted to create disfigurement of the countryside or atmospheric pollution. Many local authorities do acknowledge some responsibility in this and are prepared to make separate collections of large items if special payment is made. It is debatable whether such a charge is justified for what should be an automatic service in an enlightened community, which pays rates to ensure that towns and counties provide the facilities required by the populace.

Unfortunately arrangements for the disposal of waste even in areas that do provide suitable sites are rarely publicised. So that newcomers to a district may not know that a local dump exists, may have no idea whether the local council collects large items free or by charging, or if it is capable of disposing of old cars. There should be some co-operation between local councils, house agents and housing authorities to ensure that all new householders are informed of what is done for their benefit. It is, after all, in the interests of the community as a whole. Councils which do provide good dumping facilities often find that people will travel considerable distances from other areas with bulky rubbish because their own authorities have ignored the needs of the time.

Dealing with worn-out motor vehicles is becoming a major problem with which many small authorities are unable to cope, because of the expense of purchasing the large equipment required for quick crushing. However, there is now an obligation on local councils to provide dumping sites at which vehicles can be abandoned, and no excuse exists for people dumping cars along the roadside. It has been estimated that the number of redundant vehicles in Great Britain now exceeds one million per annum. Fortunately the majority of these are scrapped by reputable dealers who carry out a very necessary job. Local authorities are receiving increasing numbers of vehicles to deal with, and in 1967 were responsible for disposing of 15 per cent of the nation's discarded vehicles. Despite this and the much tighter dumping law, it was forecast that 150,000 cars would still be dumped by the roadside during 1970.

In the U.S.A. there are about seven million redundant motor vehicles every year. Although the majority of these are salvaged and cannibalised in some semi-organised manner, it has been estimated that the number of abandoned vehicles throughout the fifty States could be greater than sixteen million. Considerable emphasis is now being placed on improved methods for scrapping vehicles and for separating the ferrous and non-ferrous metals from each other.

Until recently British laws on dumping were very weak indeed. Under the Removal of Vehicles (England and Wales) Regulations, 1961, local councils were given the authority to take steps to remove any vehicle that had been abandoned by the roadside. There was, however, no obligation on them to carry out the removal. The effect of the Act was partially lost because under the Law any vehicle removed had to be retained for six weeks and could only be disposed of after a serious attempt had been made to trace the owner. Needless to say, councils were loath to give storage space to car wrecks which were also going to necessitate office time being spent checking ownership records.

The Civic Amenities Act, 1967, has much more power to deal with abandoned vehicles. Apart from having to supply dumping sites, the local authority now has a duty to remove any vehicle believed to be abandoned. This process is carried out by affixing a notice to the vehicle stating that it will be removed after seven days if the owner has failed to remove it in the meantime. In addition to this, the penalty for any traced dumper of vehicles or other rubbish is a fine of £100 for conviction of a first offence. Despite its good intentions and increased powers, it is still possible to see vehicles which have been abandoned rusting by the roadside for weeks and even months before removal takes place. Also the Act seems to have been completely ineffective in controlling the itinerant caravan dweller who exists by scrapping cars on the grass verge and then moving on when the created squalor is even more than he can bear.

There are two additional steps that can be taken to overcome this type of situation. The first is to improve the nature of vehicle registration so that the last owner is automatically responsible for the eventual authorised disposal of the vehicle. This would necessitate making the car log book more comprehensive, which would have the added advantage of preventing the 'ringing' and sale of stolen vehicles. Former owners of vehicles not being road-taxed would have to supply a certificate of scrapping from a registered scrap dealer or local authority or evidence of transfer of ownership. These controls should be relatively easy to operate once car vehicle registration has been centralised and fully computerised.

The alternative suggestion that has been made is that every new vehicle when sold should include in its price a returnable scrapping fee, which would be repaid to the owner, who eventually decides that the vehicle is no longer roadworthy. Objections to this are on the grounds that the Government has an interest-free loan for the lifetime of the car and that the person obtaining the refund is not the original purchaser of the vehicle. Both points are minor objections, as the scrapping fee could be registered in Government bonds with a built-in interest, and the vehicle would carry the refund fee value in its secondhand price at all stages. No doubt the Civic Amenities Act will need to operate for about five years for its effects to be fully measured. Further tightening up on litter and dumping will only come about as public pressure on the matter grows.

As the laws on waste disposal now stand, local authorities are not obliged to give the same service to shops and industrial premises as they do to private householders, despite the considerable sums paid by them into the local exchequers. The position laid down by the 1936 Public Health Act entitles a local authority to decide for itself whether or not it will remove trade and industrial waste. If it chooses to do so, it is obliged to make some additional charge for this service – this is on top of the portion of the local rates already paid by the shop or manufacturer which goes towards providing a waste-disposal service in the area. Most local authorities do, in fact, collect the refuse from shops by special arrangements with the owners of the premises, and some will take certain specified wastes from industrial plants. There is still, however, an obligation on manufacturers to deal with their own waste disposal problems, so that a great deal of on-site incineration occurs in factories. In some cases disposal difficulties have led to fly-by-night dumping in the open countryside. North Wales has had some particularly bad examples of this in recent years with toxic chemicals being discarded in areas from which they could have been carried into local rivers used for domestic water supplies.

The difficulties of dealing with industrial wastes have grown so rapidly that in 1964 the British Government set up a special committee under the chairmanship of Dr A. Key to look into it. The report, published in 1970, dealt with the subject in considerable depth. It pointed out that most industrial solid waste is inert and is suitable for land reclamation. One of the committee's most important recommendations, to deal with fly-by-night dumping, was that all private disposal contractors should be licensed. This would mean that industry could only have its wastes dealt with by approved contractors whose methods satisfied certain defined standards. The idea has the full support of the more responsible disposal companies.

Correct disposal of industrial waste has developed into a very well-organised and sophisticated specialist undertaking, in which several large companies are now involved throughout Great Britain. In many cases there really is 'money in muck'. One company dealing in industrial waste increased its trading profit by over 250 per cent in four years, from just under £210,000 in 1965 to over £680,000 in 1969. Companies operating in this field of work have to maintain the same high level of standards as do local authorities. This means that incinerators must have adequate pollution controls, chemicals must be carefully neutralised, and tips must be filled according to recommended procedures. Any cheap-jack operator trying to cut corners will be liable to find himself up in court for causing smoke pollution or creating offensive smells – poor disposal advertises itself most effectively!

Very high standards of waste disposal already exist in Great Britain, but the magnitude of the problem is increasing. During the next few years the main task for local authorities will be to ensure that disposal capacity exceeds the demand and that the accompanying pollution control is adequate to meet tighter restrictions. At the same time private individuals will need to become much more litter-conscious and education on this must start at an early age. The Keep Britain Tidy Group has spent many years trying to improve the level of public morality in this matter. Part of the answer lies in more and better situated litter bins. Also car manufacturers could help by fitting better waste boxes in the passenger area of cars and by encouraging filling stations to provide motorists' litter bins. But obviously the problem can only finally be solved when every single individual cares deeply about keeping town and countryside clean and beautiful.

8 This green and pleasant land

There are three main methods by which any area of green hills or woods and valleys can be ruined. They can be poisoned by chemicals, pollution fallout or drainage from tips; they can be denuded of their vegetation; and they can be desecrated by excavations, waste tips and the constructions of modern civilisation. All of these are happening now and have been happening for many years. Poisoning of the land by herbicides, fungicides, insecticides and discarded chemicals has become a common occurrence. The uprooting of trees and hedges by farmers to enlarge fields and the total removal of woods by developers are happening every day in some part of the country. At the same time industry is devouring land with mammoth excavators and pouring forth its waste for storage in offensive spoil heaps. None of this is new but much of it is totally unnecessary and, fortunately, some of the damage is now being reversed.

The Jekyll and Hyde properties of pesticides have led to opposing camps facing each other in verbal argument, often throwing objectivity overboard in order to score a point. Because of the large number of chemicals employed to control pests and weeds the generalisations applicable in one case may be totally irrelevant in another. It is not simply a matter of 'they are all bad'; each individual substance must be considered on its merits after investigation into both short-term and long-term effects.

Attempts to condemn all pesticides have grown as a result

of the rise and fall in the fortunes of D.D.T. and related chemicals. Much of the condemnation has been well justified and in recent years many of the fears expressed by Rachel Carson in her book *Silent Spring* have been vindicated both in the U.S.A. and other countries. This does not mean that chemicals have no part to play in agriculture and horticulture; but it does mean that in the future more careful screening and studying of long-term effects must be carried out on any potential new pesticide, before it can be applied by an industry whose function is to provide the food which we eat.

It is worth reflecting on the past results from failure to control agricultural pests and diseases. One particular tragedy was the potato famine in Ireland. This led to a million people starving because there was no known method to control potato blight in the middle of the nineteenth century. Similarly before the advent of organic pesticides life expectancy in many under-developed countries was less than thirty years. Whilst crop sprays and chemicals have not overcome all agricultural worries or solved India and Chinas' population problems they must be given much credit for having helped enormously to increase world food supplies.

Attempts to control plant pests and diseases have been made for centuries. The Babylonians used a charm in efforts to rid themselves of the fungus disease, ergot, on rye, and the Romans used pruning knives coated with bear fat as an anti-fungal precaution. Despite these early methods many peoples believed that crop pests and diseases were either acts of God or the works of satanic forces, and not until the end of the nineteenth century were chemicals enlisted in the fight against these destroyers. These first-generation pesticides included lead arsenate, introduced in 1892 to prevent the spread of the gipsy moth, and Bordeaux mixture, a mixture of copper sulphate and lime, still used to control fungal diseases on plants today. Many of the early pesticides have now been superseded by organic compounds. In some cases this was because they led to accumulations of unwanted metals in the fields. This occurred with arsenic in which continual spraying over many years led to substantial toxic residues accumulating in the top layer of the soil.

The emphasis swung to organochloride pesticides during and after World War II, heralding the boom years of the second-generation pesticides. The attack was spearheaded by D.D.T. This was originally introduced to combat the *Anopheles* mosquito, which it did initially very effectively. D.D.T. and the other organochloride pesticides were the first of many substances, now numbering about three hundred in general use, which are active against a wide range of insects. For a short period they appeared to have created a situation in which man had become the master in the battle against insects and plant diseases.

The short-comings of this form of chemical warfare only became evident after several years, although earlier warnings were uttered by more far-sighted individuals. The pesticides that had been developed were effective because they were toxic to the organisms against which they were supposed to be active. This very fact meant that in sufficient quantities they could be toxic to birds and fish, and even to domestic pets and human beings. In theory they were not supposed to be used in concentrations likely to cause death to animals, but, as was eventually seen, this was not so easily controlled. What was forgotten in the euphoria of initial success, was that although these chemicals were active against a wide range of insects, in fact only 0·1 per cent of insects are classified as pests. As a result many harmless insects were also exterminated.

Increasing quantities of pesticides and new chemical variations were found to be necessary, as insects developed strains which were resistant to the originally highly successful organochloride compounds. With many insects evolution of resistant strains was very rapid – the *Anopheles* mosquito is now completely resistant to D.D.T. in many areas, and the plague-bearing flea in Vietnam has developed a resistance to a wide range of pesticides. As greater quantities of some of these pesticides became necessary, so the likelihood of them being used at levels toxic to wild-life increased. Results of this were rapidly seen as reports occurred of the deaths of substantial bird populations and of fresh-water fish in areas after crop spraying had been carried out.

Another danger that led to opposition to the use of these

particular pesticides was their persistence. Their chemical nature was such that they did not decompose or break down readily after their original purpose had been carried out. Their persistence in this way meant that levels of pesticides were gradually building up in the soil as spraying was continued year after year, leading to the elimination of beneficial organisms in the soil. In addition to this it meant that fruit and vegetables were being sold with pesticide residues still adhering to the skin and leaves. Because of the insoluble nature of many of these compounds, not all of these residues were removed even after further washing. In some cases the pesticides were even capable of penetrating into the edible portion of the fruit. D.D.T. appears to be by far the most persistent of all the organochloride pesticides, with very little breakdown occurring even over periods of several years. Even the less persistent pesticide Aldrin only diminishes by 50 per cent during a two-month time lapse.

Organochloride pesticides are by no means the only culprits, as accidents, over-dosing and persistence occur with many other synthetic chemicals. Even some of the so-called low-persistent organophosphorus compounds such as Parathion have 50 per cent of the original dose remaining after three weeks and take over three months to reach negligible proportions. During one twelve-month period this pesticide alone was responsible for five deaths in the tobacco-growing state of North Carolina. Generally health hazards arise from misuse and carelessness, rather than from deliberate acts, but as a result lives can be put in danger. Unfortunately few people are fully aware of the toxicity of the substances that they use, and often misuse, so freely.

Persistence of pesticides means that continual ingestion of these chemicals by animals can lead to an accumulation of the pesticides within the fat tissue of the body, a factor that has been reported in most countries throughout the world. In the U.S.A. foods containing 7 p.p.m. of D.D.T. in the fatty tissue are considered unfit for human consumption. On this basis the Americans would be advised not to become cannibals, as the average U.S. citizen has been found to have accumulated pesticide levels of 12 p.p.m. in his body tissues.

The extensive use of persistent pesticides has caused further undesirable effects through the food chains which exist of animals preying on lesser creatures. It has been found that even if D.D.T. is used in very low concentrations on plants this can be concentrated to toxic levels by higher forms of life. An example of this was found with plankton, which acquired a D.D.T. level of only 0·4 p.p.m. from pesticide residues in the Long Island estuary. These plankton are the food source for clams, which tend to retain any D.D.T., and were found to have built up a level of 0·42 p.p.m. The clams in turn are a major source of food for gulls, which in turn increase the concentration of D.D.T. to levels as high as 20 p.p.m.

The same type of concentration of residues can occur in food chains starting with land plants such as berries and fruits. The toxicity of the residues has meant that birds' eggs are laid containing pesticide levels related to that found in the adult bird. In many cases the levels have been sufficiently high to prevent eggs from hatching. This occurs either by making the eggs sterile or by leading to the formation of thin, easily broken egg shells. It is believed to have been one of the main factors in the almost total elimination of the osprey from Scotland – a situation that is now possibly being reversed by very careful protection of any nesting birds of this species. Although the use of D.D.T. has now been banned in some countries, along with other persistent pesticides, the residues already existing in many areas will continue to have a detrimental effect on the quality of the soil and animal life for many years.

Fortunately chemists, biologists and ecologists have been developing alternative methods for dealing with pests for a number of years. So there is no thought of the abandonment of large quantities of the world's food supply to predators because a temporary success turned out to be a long-term failure.

The new approach has been to produce what are being called 'third-generation insecticides', developed on an entirely different basis from the earlier chemicals and with the benefit of hindsight as to what should be avoided. Most new pesticides are, therefore, only issued after very careful screening to ensure that they act selectively against only one form, or a limited group, of pests. In addition they must be non-toxic to wild-life, degradable within a short period of time and act in a manner which prevents the development of resistent strains.

Chemicals used in the new approach to pest control are substances which either prevent the predators from reproducing, by making one of the sexes sterile, or alternatively prevent the larvae stage of the insect from attaining a mature form to complete the life cycle. Such compounds are, therefore, tailor-made to interefere with the normal processes of the selected insect, and to do this often possess chemical structures closely resembling substances produced naturally by the predator. This field of work is still in an early stage; in the meantime large quantities of 'second-generation pesticides' are still in use, and considerable care is necessary to avoid another D.D.T. situation. There is already sufficient evidence that in some places the decrease in wild-life has already produced a situation too near that envisaged in *Silent Spring*.

In various research laboratories throughout the world considerable efforts are being made to find alternatives to chemical methods of control. Biological control has been based upon two different techniques. The first of these requires the release of sterilised male insects to compete with normal males in mating with the female insects. If large enough numbers of the sterile insects are used a large proportion of eggs laid are infertile and will not hatch out. By repeating the programme over several years almost total elimination of the pest can be achieved. This approach has been particularly successful in eradicating the screw-worm fly, a parasite which can cause death to cattle, from the south-eastern part of the U.S.A. and from Texas.

The method depends upon being able to raise large quantities of the particular insects under artificial conditions and of finding a suitable method of sterilisation. Both irradiation with X-rays and chemical treatment have been found effective for this purpose. Although this approach was first proposed about thirty years ago major research into it is still a relatively new thing. Obviously it is not possible to produce overnight solutions to particular problems by this means, but in the long run wider applications of the method should become available and help to remove some of the worst pests known to man.

A second biological approach has been to find another insect that will prey upon the pest it is intended to eradicate. This method was operated with some considerable success in Australia, which was plagued by the tremendous spread of the prickly pear in various parts of the country. Introduction of a natural predator, a parasitic insect *Cactoblastis cactorum*, rapidly reversed the trend and overcame the problem. This method of control has been used in a large number of different countries, frequently with tremendous success. It has, however, always been overshadowed by the apparently overwhelming success of chemical pesticides. Even the home gardener practises this biological form of control when he carefully avoids treading on a ladybird or allows a spider to continue living. There are dangers with this approach which have to be carefully checked before a predator can be introduced. In particular there is always the possibility that if the pest is totally eradicated the introduced insect may itself become a pest and cause more damage than the creature it displaced. This can only be avoided by foresight and careful research.

Chemicals still have a role to play in pest control other than in the form of pesticides. Insects are often attracted to each other, to bring about mating, by means of sex attractants which can be detected over distances of several hundreds of yards. Synthesis of even small quantities of the chemicals which act as sex attractants can enable traps to be set which will attract all the insects of one sex of a species within an area to be caught and destroyed. Considerable research is underway to prepare such chemical attractants. This again is a very encouraging line of work and may lead to almost total elimination of some pests.

'Prevention is better than cure' is an old saying that is very applicable to land conservation and the control of pests. If the spread of a particular pest can be stopped or reversed it avoids the use of pesticides which might kill wild-life and ruin the soil. This is the thinking behind the controls which exist on the roads into the states of California and Arizona. At these boundaries travellers are stopped to ensure that they are not importing fruit from other parts of the U.S.A. which might harbour diseases or pests foreign to those two states. In a similar way the considerable controls against the Colorado Beetle in Great Britain have prevented it from being a scourge to the potato crop and means that no special pesticide sprays are required for this purpose. Of course, this method cannot always be successful, as insects and infections are no respecters of international frontiers and can be transmitted from one area to another by accident.

From this it can be seen that the farmer and horticulturist need not be solely dependent upon pesticides for the adequate control of some common plant destroyers. Farming is, however, an industry which even with all its modern equipment and methods only supplies about half the food requirements of Great Britain by home production. To improve yields the farmer must use whatever appear to be the most effective and economical techniques. It is not enough to tell him that certain pesticides are likely to kill wild-life or may drain into the rivers. What must be done is to develop and demonstrate superior alternative methods of control; or, if chemicals must still be used, to ensure that farmers and farm workers are fully informed of their dangers and the correct methods of handling them.

It has been amply demonstrated that aerial spraying is one of the most wasteful and uncontrollable methods of covering any crop but it is still used in most countries despite these disadvantages. Not until the end of 1970 did Sweden become the first nation to ban aerial spraying. In many cases more efficient coverage can be achieved with lower concentrations by spraying at ground level. This means that the land can recover quicker from any undesirable effects and there is much less likelihood of the spray drifting outside the required area.

The two differing requirements of farming and conservation necessitate a compromise at all times. This is particularly so in the increasing tendency for fields to be enlarged by the removal of hedges and trees. It has been estimated that during the past twenty-five years hedges have been uprooted at the rate of 5,000 miles every year. Removal has been carried out extensively on farms which do not raise livestock as larger fields assist more efficient and economical use of modern machinery. Effects of the removal of hedges are not fully established, but the indications are that when the length of hedges in any area falls below roughly two miles on every two acres of land it automatically leads to a reduction in the bird-life nesting in the area. At the same time hedge height can play a major part in the success of nestings, as research indicates that hedges trimmed shorter than 4 feet in height enable predators such as rats to reach easily the nests built in them.

A conflict, therefore, exists between ornithologists and country lovers on the one hand, wanting plenty of trees and hedges as natural habitats for wild-life and as essential to scenic beauty, and the farmer who needs to make his farming as productive as possible. Fortunately there is some common ground between the two. The farmer is often only too happy to retain hedges which act as boundaries to his property as these keep out trespassers more effectively than do fences. Also, hedges, although needing occasional trimming, rarely require to be replaced completely every few years and usually withstand adverse weather conditions better. Not all farmers are anxious to tear down trees to form a smooth, bleak-looking landscape. As wind-breaks and in the prevention of soil erosion they are of great importance, and no farmer wishes to see his land turned into a dust bowl. One important step in the conservation of hedges has been taken by the Crown Estate Commissioners, who now pay a special bonus to their tenant-farmers to improve and plant both hedges and trees. This policy will apply to nearly a quarter of a million acres in Britain.

Increasing demands on land owing to expansion of towns, the development of motorways and airports, and the enlarging of fields, have all contributed to a substantial reduction in trees and bushes throughout the country. The denuding of the land only occurs gradually and for the greater part passes relatively unnoticed. The increase in carbon dioxide in the atmosphere and a natural desire to maintain a pleasant landscape are two reasons why this trend needs to be reversed.

What has come to be referred to as 'planning blight' can lead to disfigurement in almost any part of the country; a nuclear power station in a national park, a motorway cutting across the Chilterns, and a reservoir flooding a beautiful valley, have all become realities in recent years. Permission for this type of development has frequently been granted despite considerable opposition; often because of the relatively limited controls which exist on statutory authorities such as the Gas and Electricity Boards. Once devastation of an area has taken place considerable efforts are necessary to reduce the effects of the development to a minimum. By good planning, landscaping and replanting even the worst siting can be improved.

When the motorways were first constructed they passed through barren cuttings which sliced through fields and woods and were bounded on either side by characterless barred fences. But the years, and a little thought, have mellowed their edges slightly. Many fences are now covered by hedges, and the grass and plants on the embankments are now allowed to grow more freely without the weekly threat of the mower to restore uniformity. Above all there is a policy of planting trees along the motorways to help hide the worst features of bridge contouring; it is only a pity that enlightenment here so rarely extends beyond geometrical regularity in regimental rows. The present measures are really only a beginning and plenty of scope still remains for improvements where the demands of modern civilisation have desecrated the landscape.

Development of roads and cities will not come to a complete halt at any time, irrespective of the strength and growth of any conservation movement. It is, therefore, necessary that further expansions are carried out with more imagination and care for co-ordination between the demands of expediency and aesthetic quality. Motorways are a case in point here. Improvements in appearance can be obtained if the two carriageways are considered as independent roads to follow the natural contours separately. This not only creates a more aesthetically pleasing layout, but overcomes the dangers of head-on collisions from cars crossing from one carriageway to the other, and helps to prevent night-time dazzle from oncoming headlamps. Also it is now the policy of the Department of the Environment to smooth off the sharp edges of motorway embankments to enable cultivation to be continued further along the smoothed slope, thus saving some of the agricultural land.

One of the most useful analytical instruments for analysis of air and water pollutants is gas chromatography. By this device minute quantities of materials can be separated from each other and their presence visually demonstrated on the chart recorder.

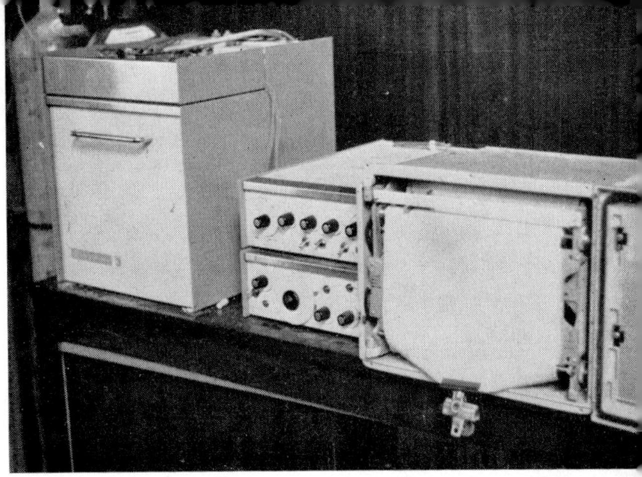

Even society's watchdogs are not immune from the polluters. The National Society for Clean Air has been campaigning against air pollution since 1899 and has played a major part in organising public opinion in favour of control legislation.

Most people are totally unaware of the vast quantities of rubbish produced by each family until some special circumstances bring it to their attention. During the 'dirty jobs' strike in 1970 emergency dumps were created everywhere – in the middle of Leicester Square, and alongside major roads such as the Sidcup by-pass.

Disposal of large quantities of domestic rubbish is necessitating modern plants such as this shown in the construction model of the Greater London Council's Edmonton Incinerator. The plant possesses twenty-three discharging bays to take the continuous flow of waste from eight of London's boroughs.

Despite tighter laws against the dumping of waste, roadside car wrecking such as this alongside the Sevenoaks, Kent, by-pass is all too common and continues with the culprits unpenalised.

The very erection of a sign in itself seems an open invitation for the countryside spoilers to move in and show their total disregard for both the law and civilised behaviour.

Similar considerations applied to housing developments, and city centres can make the difference between a town in which people just exist and one in which people can live and feel they belong. Traffic-free pedestrian precincts can be attractively laid out with trees and shrubberies, and open spaces can be landscaped into new developments. Even in old-established areas improvements are possible by demolishing some properties to create open spaces. Community schemes of this type have been carried out successfully in Leeds and Exeter for example. By closing streets to through traffic and planting trees, tedious old terraced housing estates have been turned into areas with increased character and community life.

The planting of more trees and shrubs should be carried out on a much greater scale throughout the whole country. There needs to be a positive policy of planting trees on new estates as soon as they are completed and of creating more sites in established areas. This is particularly needed in many large council estates where the only large plants are often privet hedges or some pathetic, regularly lopped plane trees standing by the kerbside. The Automobile Association showed how planting could be organised and how prepared the general public is to support such a scheme when their 'Plant a Tree' programme was introduced in 1969. But one scheme of this type is not enough! It needs repeating on ten times the scale every year. Nor is it sufficient to plant only trees which will mature within a space of three or four years. There is a need for variety in the planting. Oak trees planted now will be standing tall and strong in fifty or a hundred years' time just as we now look at those that people with foresight planted so long ago.

John Barr, in his book *Derelict Britain*, has pointed out that woodlands only cover about 7 per cent of the area of Great Britain and that this is considerably less than in most other European countries. The main authority with responsibility for afforestation is the Forestry Commission. Parliament has imposed upon the Commission the duty to develop afforestation and increase the growth and supply of timber. This means its main purpose is to look upon trees as a crop to be grown and harvested for financial return. Its added duties of providing public recreation facilities and wild-life conservation are ancillary to economic considerations. The Commission is empowered to make grants to local authorities and organisations to assist in amenity developments, but these grants can only be made if it can be shown that the long-term intention is timber production, rather than provision of recreation amenities or environmental improvement.

It is this prior consideration for timber production that many people feel limits the real effectiveness of the Forestry Commission. It means that trees are planted where they will grow most rapidly and give the fastest return on investment. But where they are needed is often in areas where the soil is of poor quality and the general growing conditions require careful treatment over a long period. It also means that there is a tendency for pines and other soft-woods to be planted in preference to the more beautiful and longer lasting hardwoods such as oak just because the former help to satisfy an ever-increasing world demand.

John Evelyn was a man of remarkable perception. Apart from his concern over London's smoky atmosphere he was also a keen advocate for the preservation of trees and forests. In 1664 he published his book *Sylva, – or a discourse of Forest-trees and the Propagation of Timber in His Majesties Dominions* in which he wrote: 'But what shall I then say of our late prodigious spoilers, whose furious devastation of so many goodly Woods and Forests have left an infamy on their names and memories not quickly to be forgotten!'

His ideas for replanting areas that had been cleared in addition to the extension of copses, borders and hedgerows are equally applicable today: we have not moved very far in three hundred years. Evelyn also maintained that it was necessary to establish a balance between the simple economic considerations of timber and purely aesthetic demands. A balance between the two is probably of even greater importance now than it was at that time, but there is undoubtedly still plenty of scope for improving the wooded areas of Britain to suit both criteria.

Whilst the Government and local authorities must play a major role in increasing and improving woodland, industry and private individuals still have a part to play. Industry can mask some of the more unpleasant aspects of factory buildings with careful landscaping coupled with judicious planting of trees and bushes. Increasing numbers of tree nurseries supplying semi-mature trees make it possible to achieve improvements in a very short space of time. Private individuals can help by ensuring that local restrictions against tree-felling are observed. They can also adopt a policy of having at least one tree in any garden, ideally coupled with a nesting box and bird table. This may seem a small thing to do but multiplied millions of times over and coupled with larger national planting and landscaping programmes would help in bringing a major improvement in the environment.

It is all too easy to leave everything to the mythical 'they' stuck in some Whitehall office to make the decisions. We may not all be able to help build a new sewage works or install an electrostatic precipitator but we can plant a tree and we can help birds and animals to survive. We can also ensure that our elected representatives are made aware that we do not object to spending the nation's (which in any case is our) money to obtain better landscapes, more national parks and an improved environment.

Fewer pesticides and more tree-planting, however, are not the only ways in which the quality of our land and scenery can be improved. All around the country we have the scars and growths of the legacy from our industrial development. Often they are just eyesores spoiling the landscape and defacing the hillsides. All too frequently they are also poisonous with their dangerous chemicals being leached out into the rivers, and occasionally they are a very great danger to people living near them. Anybody who has read Tony Austin's book of the Aberfan disaster can only be horrified at the nature of the tragedy which arose from the discarding of mining waste in spoil tips around the mountain side. In that particular disaster 144 children and adults died because of a combination of avarice, ignorance, procrastination and lack of foresight. But that was in 1966; and still tipping goes on throughout the country, often with total disregard for local considerations.

The simple answer has always been to use the waste to fill

in holes left by the excavation of sand and gravel or other minerals. Unfortunately the waste and the holes are often in widely separate parts of the country and arguments occur over who will pay for the loading and transportation. Such situations are not helped by 'mock auctions' of the type used by the B.B.C. television programme *Nationwide* in August 1970 when they set out to embarrass the china clay companies in Cornwall over the difficulties of removing the white waste heaps that spoil the landscape around St Austell.

Waste heaps that have been built up over years, whether originating from private or nationalised industries, can rarely be removed at short notice, however desirable this might be. The first thing to be done is to take steps to ensure that further dumping is only permitted on a short-term basis when it can be shown that there are eventual plans for removing and disposing of the waste in a more civilised manner. With established tips there are two main solutions: removal and landscaping.

Removal of a tip can be achieved in some cases by finding a use for the waste material. Often large quantities of solid ballast are needed for forming the foundations to motorways. Similarly some waste materials can be used for making blocks suitable for building. In certain cases it is even economic to process old waste dumps to extract metals whose prices have increased. By far the easiest, and most common, method of removal is by use in land filling and reclamation. The main problem is the cost of labour and transportation from the tip to the reclamation site. Costs involved in this form of environmental improvement will automatically be absorbed in higher prices. Even if the Government provides grants for the work to be done it must be reflected in higher taxes or local rates. So that, directly or indirectly, everyone will have to pay for such things to be carried out. It is by no means certain that the general public is currently prepared to face up to the fact that, irrespective of where the responsibility for improvement is placed, the community at large will have to bear the cost; just as they have benefited from cheaper products during the period that waste materials have been easily dumped in great heaps.

It is a mistake to assume that this problem of spoil tip removal only applies to the industrial north and Wales. Even the London area has tips which blight the landscape. In one particular case, in the London Borough of Newham, the situation is even more unjustifiable as the tip is situated alongside a large area of low-lying, poor-quality land, and only needs to be transported a few hundred yards to be usefully employed in raising the land level. Undoubtedly many similar situations occur in other parts of the country.

Often the cheapest method of improvement is that of landscaping the tip, covering it with top-soil, and planting grass and trees. This means spreading the waste out over a larger area than was originally used in order to round it off to fit in with the general contours of the area. By this means it is possible to create new grazing land for farm animals, playing fields in mining towns, and new wooded areas. There have been many successes in dealing with spoil tips and, given the desire, inspiration and money, there is little reason why the nation should continue to suffer from the many eyesores still covering the land. It is, however, becoming evident that greater Government intervention will be necessary in this aspect of environmental improvement, both to ensure that some of the necessary money is available, and to cajole and persuade interested parties to get together and make the arrangements to deal with the situation.

Once more, even in this field of planning and improvement, there is a part that the general public have played and can continue to do so with even greater effect in the future. By keeping a vigilant watch over planning and development proposals many local groups have managed to get prospective eyesores eliminated before they have even left the drawing board. An example of this occurred in 1966, when the Southern Gas Board announced its intentions of siting a large conventional style gas-holder in Abingdon to tower 128 feet above the town. The local council was only informed by the Gas Board out of courtesy. When objections were forthcoming from practically every interested body in Abingdon the Gas Board then tried to persuade the local council into paying the £300,000 additional expenditure for a high-pressure gas-holder that could be fairly well hidden. Eventually the Gas Board settled for the local authority paying £120,000 of the additional bill. In fact the rest of the country owes a debt to the people of Abingdon for the fight they put up because, as a result of the publicity this case received, the House of Lords introduced an amendment into the 1968 Town and Country Planning Act that makes it impossible for a local authority to be held to this type of blackmail resulting from the refusal to grant planning consent.

Success is not always achieved, however: the battle for Bacton, the Norfolk coast site for North Sea gas, was won by the Gas Board. The route of the M40 through the Chilterns has been decided (but may yet be revoked) against the alternative route proposed by the amenity societies in the area. Public concern has shown itself and been clearly expressed in these cases, and has helped to make more people aware of the way in which public bodies and statutory undertakings often attempt to bulldoze proposals through the planning stages.

Organised opposition costs time and money, but this is the essence of democracy. Examples of this include the tremendous campaigns fought by the areas considered as potential sites for London's third airport. At Wing alone, 75,000 signatures and £30,000 were raised in one year, enabling experts to be engaged and full legal representation to be obtained. Without the opposition and concern shown by this type of action, not only our countryside but our rights also will be lost. When irresponsible, inconsiderate, selfish development and waste-tipping is allowed to occur unopposed, we shall all be a step nearer to the Government dictatorship that the majority of people oppose.

It is often thought that present laws to prevent planning blight and improve public amenities are inadequate, and no doubt further regulations will be introduced as the nation becomes more dissatisfied with continued disfigurement of the land. For all that, fairly substantial powers already exist to do much of what is required. For instance, under the National Parks and Access to the Countryside Act, 1949, every local authority has the power to create nature reserves in which wild-life can be protected and assisted in their battle for survival. But how many borough councils or even county councils have used the powers they have been given? In the case of waste tips, new tips can only be created with the approval of local authorities. The weakness in the law here is that tips established prior to 1948 can continue to be used without further approval being required. One can envisage this situation leading to an old tip growing to gigantic proportions because of the loophole.

Attempts have been made to justify the special privileges given to the statutory undertakings such as the Electricity and Gas Boards, on the grounds that, as they are charged with a statutory duty to supply particular services in an area, they should not be prevented from carrying out this duty by local authorities withholding the necessary planning consent. Because of this, the various Acts of Parliament creating these bodies, and the Town and Country Planning Acts since 1947 have established procedures by which the normal planning steps can be by-passed. Even the 1968 Act still permits this despite the increasing opposition from local authorities who resent being faced with a *fait accompli* owing to these bodies being able to go above their heads.

Fortunately there are now substantial forces allied in the fight for the environment and for conservation. Apart from Government bodies such as the Countryside Commission, which administers the National Parks, and the Nature Conservancy, which runs nature reserves and carries out research on conservation and wild-life, there is a large number of independent bodies dealing with the many different aspects of nature and the countryside. Most of these organisations operate with no form of Government support or subsidy. The extent of the increase in public concern for the quality of the countryside and living conditions is indicated by the fact that in recent years most of these bodies have reported substantial increases in membership.

The National Trust is probably the best known of such organisations and was founded in 1895. Its annual income is of the order of £2 million from subscriptions, bequests, donations and entrance fees. During its existence it has acquired ownership and protection of over 400,000 acres of land in addition to over one hundred historic houses and monuments. Under the National Trust Act, 1907, property owned by the Trust has been declared inalienable and cannot be disposed of or compulsorily acquired without the consent of Parliament. It therefore has tremendous protection over its property greatly exceeding almost anything else in the nation.

One of the Trust's greatest ventures was its 'Enterprise Neptune', launched in May 1965. This was designed to focus attention on the depletion and deterioration of the coastline owing to continual development. Its objective was to acquire as much of the unspoilt coastline for preservation purposes as was possible. For this purpose a target of £2 million was set; the first million was reached in the remarkably short time of two years. As a result of 'Enterprise Neptune' the Trust now owns or controls over 300 miles of coastline (about 10 per cent of the total) in England, Wales and Ulster (a separate Trust covers Scotland).

Typical of the properties obtained are the 2 miles of coast at Woodspring Priory in Somerset, only 5 miles north of Weston-super-Mare, and the sand dunes and pine woods at Formby Point only a few miles from Liverpool. As a result of the efforts of the National Trust and Governmental bodies there is now some measure of protection covering 50 per cent of the coastline in England and Wales. At the same time there remain 200 miles of coastline which are still controlled by the Ministry of Defence and to which the general public have virtually no access.

About the oldest of the wild-life protection bodies is the Royal Society for the Protection of Birds, founded in 1889. It now has a membership of over 50,000, plus an annual income of £200,000. The Society manages thirty reserves, including one at Teesside in the middle of substantial industrial development. Probably the greatest success achieved by the Society in recent years has been the protection of the nesting ospreys at Loch Garten. By 1969 there were three pairs of birds each of which successfully raised two young under the binoculars of nearly 40,000 visitors. The headquarters of the R.S.P.B. at the Lodge, Sandy, Bedfordshire include a Nature Discovery Room specially designed with school parties in mind, as well as a nature trail and lecture rooms. So far the Society has failed in its attempt to obtain a total ban on the organochloride pesticide D.D.T.

Another very active independent body is the Council for the Protection of Rural England (there are corresponding Councils operating in Wales and Scotland). This was founded in 1926 with the aim of making people aware of the dangers of development encroaching upon the countryside. The C.P.R.E. receives no Government grant and is financed by subscriptions from private individuals and organisations. It serves to co-ordinate the efforts of local protection and conservation bodies and acts as a centre for the dissemination of information. It also co-operates closely with other environmental organisations and has played a major part in fighting many of the more blatant attempts at ruining the remaining attractive parts of the country. The C.P.R.E. is now working closely with the Forestry Commission over the location of reafforestation and access for the general public for recreational purposes.

One other body that deserves a mention is the Civic Trust, started as recently as 1957. Its purpose is to increase civic pride by encouraging higher standards in planning and design and to protect the countryside. One of its first successes was the improvement of Magdalen Street, Norwich, achieved by a combination of repainting and removal of unnecessary street furniture. The Trust has encouraged attempts at large-scale tree-planting and, at the request of the local authorities in the area, carried out the viability study on the Lee Valley Regional Park in 1964. The results of its efforts and recommendations for the Lee valley are beginning to bear fruit as the first stages of this 'lung for London' are now coming into existence. For an organisation that has had such a short life the Civic Trust has already made its presence felt in many different spheres of influence. Civic Trust awards for design are greatly coveted by architects and builders. The pioneering work of the Trust in transplanting semi-mature trees to housing estates, clearing litter, and tidying up drab streets, has acted as a spur and inspiration to many authorities. It is proof indeed that the spoilers are not getting it all their own way.

Great Britain is already one of the most crowded countries

in the world. The total land area works out to less than 1 acre per head of population and only about half of that is available for agricultural purposes. In the U.S.A. there is about ten times that amount of land per person, although much of this is also unusable. In all countries developing industry, increasing populations and higher standards of living have put tremendous pressures on unexploited areas of land. As the population increases so does the demand for more houses, schools, factories and roads. At the same time more people require improved leisure facilities of all forms as their free time becomes more plentiful.

The conservationists are not just fighting a rearguard action; they are trying to keep the good parts of our heritage for posterity and simultaneously attempting to improve the remainder. As the pressure for land becomes even greater the battles are going to become fiercer. It will become even more important to ensure that derelict land is improved and that adequate protection is given to the remaining beautiful areas of the countryside. At the same time as more and more people will be living in large cities it is essential that as much time and effort is spent on making city environments pleasant as is devoted to providing adequate recreational facilities outside. With many conflicting interests to be reconciled it is far from easy to satisfy everyone, particularly if population predictions of 66 millions for Great Britain by the end of the century are correct.

Land is very precious; we cannot afford to ruin it with poisonous chemicals, to scar it with excavations and tips or to cover it with vast expanses of concrete in the careless manner which has prevailed for so long. Substantial attempts have already been successfully made in preserving and improving the best parts of the country. Neither the National Trust nor the other organisations can be accused of complacency in their efforts to leave something worth while for future generations. But for all their work we are still in danger of creating such an awful environment for our children and grand-children that they may yet look upon us as the generation that had the last chance to take the necessary action and were too selfish and narrow-minded to do it.

9 The unseen pollution

Of all the forms of pollution which can occur, that of radioactivity is by far the most insidious. It is invisible, tasteless and odourless, and people exposed to its dangers can remain unaware of it until a long time afterwards. The effects of even small quantities of radiation are very long term, and because of this traditional methods of waste disposal are not directly applicable to radioactive materials without special precautions being taken.

It is almost impossible to appreciate that atomic radiation was totally unknown until the end of the nineteenth century. It was not until 1896 that the French scientist, Henri Becquerel, discovered that uranium gave off radiations very similar to X-rays, which had been discovered in the previous year by Wilhelm von Röntgen. Since that time the radiations from uranium and many other substances have been closely studied and their potentialities for good and evil both fully realised. We now tend to think of nuclear power and radiation in terms of power stations, atomic bombs and the treatment of cancer. To this list the scientist and engineer would also add the isotopes for specialist investigations.

In all countries in which radioactive materials are used there are considerable restrictions and controls on the ways in which they can be handled and disposed. These restrictions have arisen from the observations of the early investigators that exposure to excessive radiation can lead to the formation of cancer in the body; a fact borne out by the prevalence of lung cancer amongst workers in uranium mines. Effects of radiation on human beings have been fully observed in the two atomic explosions in 1945 at Hiroshima and Nagasaki, and from the fallout of the first hydrogen bomb test which affected the crew of the Japanese fishing vessel *Lucky Dragon* in 1954.

Radioactivity arises from the ability of a number of the chemical elements to break down automatically to give smaller atoms, and in so doing release electrons and rays which can penetrate through even dense materials. The chemical element uranium, for example, is the heaviest naturally occurring element known to man. It actually consists of two types of uranium differing very slightly in their masses. These two forms are called isotopes, in this case designated as Uranium-235 and Uranium-238, the numbers showing their relative masses with respect to the hydrogen atom. Natural uranium found as an ore in the Belgian Congo and Australia, as well as other places, consists of 99·3 per cent Uranium-238. The remaining 0·7 per cent, however, is the lighter isotope Uranium-235, and it was this particular substance that was used to develop the first atomic bombs. Since the 1940's many other isotopes have been used for nuclear explosions and to create heat in nuclear reactors. Nearly all chemical elements exist in several different forms: thus there are four different isotopes of lead and seven isotopes of mercury. Many of these isotopes are naturally radioactive and can cause damage to living things. In fact there is a small quantity of naturally occurring radioactivity everywhere in the world, including the human body, and we are exposed to a minute amount of radiation all our lives. Dangers occur when the radiation is allowed to be concentrated in small areas and permitted to persist for extended periods of time. In effect this is what occurs with atomic and nuclear explosions. Such concentrations have been achieved by separating the radioactive isotopes such as Uranium-235 from the bulk of the inactive isotopes. This procedure is very tedious and involves numerous expensive steps which achieve progressive enrichment of the radioactive isotope. When the concentration of the radioactive isotope achieves a particular level it is termed to be 'critical', and at this stage it is producing enough rays to explode spontaneously, at the same time generating a large quantity of heat. It will be appreciated that the first of these effects is made use of in the atomic bomb, and the second is needed for the atomic reactor. In the second case some control over the reaction is necessary for it to be usefully harnessed, and this is brought about by pushing graphite rods between the enriched uranium in order to absorb the radiation being produced.

Apart from producing heat and causing cell damage to unprotected living things which are nearby, this radiation can be used to make normally stable chemical elements radioactive. This is achieved by normal non-active elements absorbing the radioactivity into their structures. So if a piece of cobalt metal is suspended in the atomic pile for a short time, it will be radioactive when removed. The radiation, in fact, consists of parts of the disintegrating element, and if some of these happen to strike another isotope the two can often unite to form a new element. This process has been artificially encouraged, with the result that parts of the uranium atom have been used as building blocks to create heavier, more complex artificial elements. Thus the elements plutonium and neptunium are both man-made, as a result of the developments in the field of radioactivity and nuclear physics. Altogether there are ninety-two naturally occurring chemical elements, to which have now been added eleven man-made elements.

It will be appreciated that if radioactive materials are continuously breaking up then the amount of radioactivity in a particular substance is gradually decreasing. The degree of persistence of radiation from a particular isotope is expressed in the length of time it takes for the radiation to drop to half its original level. This value is a constant for any weight of that isotope and is referred to as the 'half-life'. Values for half-lives vary greatly between different isotopes; for radioactive potassium the half-life is $4{\cdot}5 \times 10^8$ years, whilst for one of the isotopes of sodium it is only 14·8 hours. There is then a considerable difference between the inherent dangers of isotopes, depending upon the length of their half-life. Unfortunately the dangers do not stop there, as, although one radioactive substance will progressively decay, it may at the same time induce radioactivity in another element around it.

The true purpose of radiation protection is to ensure that people are protected at all times from exposure to levels of radiation greater than normal. This means that disposal of radioactive wastes must take a particular form as radioactivity cannot be destroyed just by incineration or by burying in the ground. It might be dispersed in this way, but the total amount of radioactivity will not be reduced in the slightest. Similarly it is necessary to ensure that emissions and effluent from atomic power stations are purified to remove any radioactive contamination which they may have picked up. Despite the very substantial controls and regulations which exist accidents have occurred and substantial quantities of radioactivity have escaped at various times.

Atomic radiation is of three types: alpha rays, which can only travel a couple of inches in air; beta rays, which will penetrate about one quarter of an inch of aluminium; and gamma rays, capable of passing through iron 10 inches thick or lead 3 inches thick. Gamma rays are particularly dangerous because of their great penetrating power, and precautionary measures must be especially rigorous to guard against them.

For safety levels of radiation to be defined it has been necessary to establish accurate methods of measuring the activity of the materials used. Two scientific units have been accepted as the basis for these measurements. These are the Curie, the amount of material containing $3{\cdot}4 \times 10^{10}$ disintegrating atoms per second; and the Röntgen, which is the amount of radiation producing $2{\cdot}1 \times 10^{9}$ pairs of ions in a volume of 1 cubic centimetre of air.

These scientific terms refer to the total amount of radioactivity emanating from an active source. From a biological and physiological point of view it is better to measure the activity in terms of the radiation dosage actually received by a subject. For this purpose the unit currently used is the 'rem'.*

* The 'rem' is a measure of the radiation penetrating the body and allows for the different destructive powers of the nuclear rays. A related term called the 'rad' does not allow for the biologically different effects. On the 'rem' basis an alpha ray is ten times more active than is a beta or a gamma ray. Because of the considerable confusion that

Under normal circumstances the human body receives in the order of 0·1 rem per annum from natural radiation sources. So that in a lifetime a total dose of about 7 rem can be expected by the average person. It should be remembered that there is no such thing as a safe level of radiation other than absolute zero. All radiation is capable of causing harm to the body and the reproductive system is particularly sensitive to the various forms of radiation to different degrees. At the same time there is a considerable difference between exposure of the complete body to radiation and a limited exposure during a teeth X-ray. Permissible levels adopted for workers in radiation fields are limited to 5 rem per annum. This corresponds to about one-fortieth of the level that would be expected to produce death in the most sensitive person. At the same time it is about fifty times the level that would be expected from background radiation.

As the dangers from radiation have become more precisely known so the maximum permissible dose for workers has been progressively reduced, as shown by the following list of permitted levels based upon figures published by the International Commission on Radiological Protection.

1934	0·2 rem per day
1936	0·1 rem per day
1950	0·3 rem per week
1956	0·1 rem per week

These figures may seem to err considerably on the side of caution but need to be looked at in the light of the fact that radiation effects are cumulative. Exposure to doses above the recommended level can quickly lead to clinically detectable blood changes occurring. Medical studies indicate that the

---

arises from using several different units it has been considered desirable to standardise on the 'rem' throughout this chapter. All dosages and exposure levels have therefore been translated as well as possible into this unit. This may not be entirely accurate from a scientific point of view but it assists clarity and comprehension.

effects from different short-term dosage levels are as follows:

0 – 25	rem	no obvious injury
25 – 50	rem	observable changes in white blood cells
50 – 100	rem	damage to blood cells
100 – 150	rem	non-lethal limit, sterility
200+	rem	some fatalities (2%), serious injury (50%)
300+	rem	fatalities 20–25%
400+	rem	fatalities 50%
600+	rem	lethal for everyone within one month

These figures are not just informed guesses; they have been established by examination of radiation victims, some of whom died as a result of their exposures.

It should be appreciated that many people are now exposed to levels of radiation much higher than the background level that normally exists. This is because the environment has been contaminated due to radioactive fallout from atomic explosions in addition to a tremendous increase in the use of radiation sources in many realms of medicine, science and engineering. Further dangers can arise from the discharge of radioactive waste into the atmosphere, the rivers and the sea; as well as from material which is spilt during use or transportation from one part of the world to another.

As many of the early investigators suffered from radiation damage the measures now taken to protect both the workers in the radiation field and the general public are considerable. It will be up to the reader to decide whether or not these are adequate and whether further nuclear power stations and reactors should be developed. The purpose of this book is to show what is being done and to try and draw attention to the limitations and to mistakes which have occurred.

In Great Britain the accumulation and disposal of radioactive wastes is strictly controlled and regulated under various Acts of Parliament. The first of these was the Atomic Energy Act, 1946, which led to the initiation of the nuclear power programme. This was followed by the Radioactive Substances Act, 1948, the Atomic Energy Authority Act, 1954, the Radioactive Substances Act, 1960, and the Nuclear Installations Act, 1965.

In the U.S.A. legislation was started with the Atomic Energy Act, 1946, and extended in 1954, so that effectively in both the United Kingdom and the U.S.A. responsibility for control of radioactive waste is vested in the governments.

The long-term disposal of any radioactive waste substance is by natural decay processes. In cases in which the half-life is only a matter of hours, or possibly days, the radioactivity will disappear during a relatively short storage period in protected surroundings. This is not often practicable, as the wastes usually consist of a mixture of isotopes with differing half-lives. Disposal methods are, therefore, normally dependent upon either being able to dilute down the waste to a safe radioactive level in order that it can be discharged into the air or the river, or alternatively to concentrating and storing it in a protective jacket, either in disused mines or at the bottom of the ocean.

The quantity of material to be stored, dumped or otherwise dispersed is substantial and increasing every year. Harwell alone produces 30 tons of radioactive solid waste every week, but this is minute compared with the amount produced in other atomic reactors. In the U.S.A. over 75 million gallons of liquid waste are now stored in large tanks because the activity is too high for any other safe method of disposal to be applied. Every year, as more nations build nuclear power stations, the problem grows and the amount of concentrated radioactivity increases.

By far the greatest sources of radioactive wastes are nuclear power reactors. The exhausted uranium from the nuclear reactions contains about thirty different radioactive elements, resulting from the fission processes which have taken place in the reactor. The solid rods of waste material are dissolved in nitric acid to form metal nitrates which are extracted by solvents. These are then permanently stored in steel tanks reinforced with concrete. Alternatively, if the bulk volume is reduced by evaporation of the solvent, storage can be carried out in more expensive stainless steel tanks. Treatment and storage in this way is costly and is avoided if other methods can be used with at least equal safety.

Another procedure which has been applied to concentrated

quantities of solid wastes is by fusing into insoluble blocks. Radioactive materials are separated from the decay products of the nuclear reaction and precipitated out of solution. The wet precipitate is mixed with sand or a similar silicaceous substance and heated to over 1,000° C to make the mixture melt. On cooling, the viscous liquid solidifies into glassy blocks which are stored at great depth in the earth or at the bottom of the sea. This method is not readily suitable for volatile decay products as these would require additional measures to prevent their escape into the atmosphere.

If liquid and soluble wastes are diluted with a sufficiently large volume of water the overall level of radioactivity of the resulting solution need be little more than the normal background value associated with that particular area. This is the reason that most nations allow certain quantities of low-activity wastes to be discharged into streams and esturies. For example, in the U.S.A. the Oak Ridge National Laboratory in Tennessee discharges waste containing Strontium-90 and Caesium-137 into the Clinch River. Similarly, in Britain, Harwell is permitted to discharge waste into the Thames and the plutonium reactor at Windscale pipes waste into the Irish Sea.

It would be wrong to believe that Harwell is deliberately disposing of vast quantities of dangerous isotopes that are likely to cause leukaemia or cancer in the inhabitants of London. In fact the effluent is given very substantial treatment to remove the majority of the radioactive substances that can be recovered. The resulting waste water has a radioactive level only slightly higher than the normal background level for ordinary water. Full sewage treatment involving settling, primary treatment, and activated sludge treatment is first carried out; and this is followed by treatment with coagulants, further settling and purification through ion-exchange columns before discharge to the river is permitted. This type of process is standard procedure for disposal of liquid wastes in a number of countries using nuclear power.

Although this is accepted practice, disposal in this manner requires very careful control and considerable knowledge of the tidal flow in estuaries and the ultimate fate of water in the river. Currently Great Britain is using the sea as a dumping ground for about 5,000 tons of low-activity waste every year. This is dropped to a depth of 100 fathoms on the edge of the continental shelf. Deep-sea dumping at depths of over 1,000 fathoms in the Atlantic is carried out by both the U.S.A. and Great Britain. Areas selected for this are intended to be away from shipping lanes and fishing grounds as they are used for high-activity wastes.

Containers used for deep-sea dumping are designed to withstand the hydrostatic pressures which exist at great depths. It is not the intention that the radioactivity should be able to escape into the ocean but should be contained until well after any radioactivity has decayed away. Frequent monitoring of the radiation in the sea dumping sites is carried out to check whether or not unintentional leakages have occurred. However, it is difficult to see what could be done to control a leak occurring at 1,000 fathoms, and the lifetime of the containers can only be a matter for conjecture. Concern over the quantity of radioactive waste discharged into the sea has been growing for many years, and it seems very likely that international agreements will soon stop the indiscriminate dumping that many countries have been carrying out.

A problem frequently encountered with radioactivity is that a large quantity of conventional waste material may be unsuitable for normal disposal methods owing to the presence of a minute amount of radioactive substances. Rather than have to store the complete bulk of the waste, various separative techniques are used to remove the active components in order that conventional disposal can be used for the remaining waste. Ion-exchange resins of the type used for water softening have been found to be particularly suitable for this purpose. The liquid waste is run down a column of the ion-exchange resin and in the process a chemical reaction occurs on the surface of the resin by which the radioactive elements in the solution are exchanged for non-active elements from the resin. The resins are capable of coping with large volumes of solution before becoming saturated with radioactivity. This method has been of particular value in removing Strontium-90 by the use of the ion-exchange processs carried out on columns of naturally occurring aluminosilicates.

One method of disposal that is no longer used is that of soil percolation by tipping liquid waste into seepage sites. This was used as a means of dealing with low-activity material but it had many disadvantages. The main danger came from the possibilities that underground drainage could lead to contamination of water sources and even to the build-up of more substantial concentrations in localised areas. This has not stopped the disposal at lower levels in the earth's crust and the use of disused mines for this purpose is common practice in most countries.

Incineration of waste materials is extensively practised where the bulk volume is likely to be substantially reduced and the radioactivity retained in the ash or clinker left in the incinerator. In cases in which this is used, very special precautions are taken to prevent the escape of radioactivity in the effluent gases. Filters, scrubbers and electrostatic precipitators are all used for this purpose. It is, however, not possible to remove 100 per cent of any escaping materials, any more than it is possible to remove 100 per cent of conventional air pollutants. Because of this, special limits on the effluents from smoke-stacks at nuclear power stations operate and what are termed 'permissible releases' only are permitted at any time. This really means that it is accepted that atomic reactors must discharge some radioactive waste into the atmosphere and it is a mistake to talk in terms of such places being constructed with total removal of all dangerous substances.

One method of disposal that has frequently been suggested, but not yet tried, is that of shooting radioactive wastes into outer space. Public objections to other forms of disposal would have to be extremely powerful to necessitate such an expensive method of dealing with the problem, and it would probably be cheaper to turn to other methods for generating electricity rather than have to fire off a multi-stage rocket once a week.

It appears that present methods of disposal are both adequate and suitable for the quantities of radioactive wastes currently being produced. But the number of atomic power stations and countries joining the nuclear group is increasing every year. The possibility that some nation might try fly-by-night dumping in an exposed site, or easy disposal into a fast-flowing river is a nightmare that could all too easily come true without stronger international supervision in this field.

Despite all the preventive measures, accidents do happen and large quantities of radioactivity have inadvertently been spread in some areas. Contrary to popular opinion, internationally the number of accidents involving radioactive materials is quite high. In his book *Reactor Safeguards*, published in 1962, C. R. Russell lists twenty-seven accidents. Some of these involved release of radiation to surrounding countryside, several caused the death of operators and some led to workers receiving heavy doses of radiation likely to lead to body damage. These only represented a small number of the accidents recorded at that time.

Accidents can be classified in three main groups: those arising from a malfunctioning in an atomic reactor used for making isotopes or to generate heat for electricity; those arising during transportation of radioactive materials; and accidents during the use of radioactive isotopes in the laboratory, hospital or factory.

By far the most dangerous are those in nuclear reactors, and on several occasions throughout the world reactors have come near to getting out of control. Although the nuclear reaction has in all cases been eventually controlled this has often only been achieved after quantities of radiation has escaped outside the reactor area. One escape of this nature is reported to have occurred at the Enrico Fermi Power Plant at Lagoona Beach, Michigan, in October 1966 during a particularly troublesome period when the whole plant came near to exploding. A similar situation had occurred in Great Britain nine years earlier when, owing to a failure of the radioactivity-arresting systems at Windscale, fission products were released into the atmosphere from the No. 1 pile. The radioactive fallout, which was calculated to be greater than that released by an atom bomb, covered an area of 400 square miles, mainly south-east of Windscale. The greatest danger came from the possibility of cows eating grass that was covered with the radioactive isotope Iodine-131, as this can pass into milk and eventually be incorporated into the thyroid gland of human beings. Because of the hazard it was necessary for the Atomic Energy Authority to impound all milk and crops in the area.

In September 1970 another accident occurred at Windscale which necessitated the closure of a large part of the plant. No reports of escape of radioactivity to the atmosphere were made, but as the original press release of the accident was not issued for a week it may be necessary to await the report of the official enquiry to find out how near the situation came to being a repeat performance of the 1957 danger. Despite the soothing tones of those who work in this realm of science, on the basis of a major accident every thirteen years in Great Britain, and more frequently on a world scale, there is a strong indication that nuclear reactors are not as safe as they have been made out to be! As it is, when the lifetime of the present reactors is ended (anticipated to be between twenty and forty years) it will not be possible to demolish them and re-use the land for long periods because of the radioactivity that has built up in the site. So by the end of the century there could be at least a dozen abandoned sites in Great Britain, along with, of course, a comparable number of new sites, probably in national parks, established in the meantime.

Other accidents are always likely during the transportation of radioactive materials, particularly as isotopes are used on a large scale throughout the world and involve carriage by all normal means of transport. Regulations cover the shape and nature of packaging required for all the radio-isotopes employed. Under normal conditions no danger is possible from the carriage of these materials throughout the world. Dangers only arise in the case of accidents, and even then the packaging is designed to absorb any radioactive materials released from its container. International regulations on packaging are very strict, and all but weakly radioactive materials must be packed securely enough to prevent leakage after being dropped 30 feet on to concrete followed by a fire at 800° C for half an hour. The strength of the radioactivity inside the package is indicated by a combination of colour coding and red stripes, and all packages must carry the well-known black and yellow sign for radiation danger. These signs serve as a guide to decontamination experts at an accident giving them some idea of the potential nature of the hazard.

From this it will be appreciated that anyone foolish enough to steal any container with a radiation symbol on it is putting himself in considerable danger whatever the value of the substance inside. Richard Curtis and Elizabeth Hogan, in their book *Perils of the Peaceful Atom,* tell the story of a small boy who found a small container of radioactive cobalt on a dump in Mexico City. As a result of someone's carelessness the boy, his sister, mother and grandmother all died from radiation exposure over a period of six months. The very nature of radioactivity makes it difficult for the general public to appreciate that there can be considerable danger in something which appears perfectly innocent. For this reason people who have to handle isotopes have a particular responsibility not to leave them around where people will pick them up or even steal them. This is not just in the interests of any thief, but for the public at large. Adequate control should mean that loss from an unattended car is not possible.

Accidents from spillage and fires in laboratories are normally localised and quite easily dealt with. To assist decontamination, laboratories which deal regularly with radioactive substances are specially designed and constructed with tiled walls, coated concrete surfaces and smooth bench worktops to enable any chemicals to be readily mopped up. Such places are equipped with counting devices which are used to check both the background count on a permanent basis and to monitor any localised concentrations of radiation. Most universities and polytechnics now use isotopes in a wide variety of ways, and personnel and students involved with this work have to undergo special instruction before any practical work can be carried out. In all establishments the use of radioactive materials is subject to Ministerial approval, which is only given if it is considered that there are adequate facilities available for storing and disposing of the radioactivity in addition to responsible staff capable of dealing with any emergency situation.

With the increase in the use and disposal of radioactive substances there is always the possibility that, although no large concentrations of radiation are created by man, they may in fact build up by other means. As has already been seen in Chapter 8, living organisms have the ability to retain certain insecticides within their cells and possibly create toxic levels. A similar but more natural situation can occur with radioactive isotopes, as plants and animals will utilise any elements, radioactive or not, that play a normal role in their metabolism. In some cases they will also utilise elements which are similar to those normally required. It is for this reason that the radioactive isotope Strontium-90 is considered to be so dangerous. It resembles calcium in the manner of its chemical action, and can be incorporated into the bone structure of all animals including human beings. Because its effective half-life is 10 years it is considered to be very toxic from a radiation point of view. The half-life value means that any infant incorporating Strontium-90 in its bones now will still retain 3 per cent of the radioactivity at the age of 60, if fortunate enough to survive that long.

This danger of incorporation has been demonstrated to occur in several cases. At the Oak Ridge National Laboratories, Tennessee, initial dilution of radioactive wastes occurs in the White Oak Lake before passing to the Clinch River. Fish taken from the lake have been shown to have built up levels of the radio-isotopes of caesium, strontium and phosphorus. The uptake of radioactive Phosphorus-32 is in fact a common occurrence among algae and other water organisms, and can act as the first step in a life chain. A study on the Columbia River, which takes water waste from the Hanford nuclear reactor, showed that insignificant amounts of radioactivity were greatly concentrated through the various steps of the life chain. Eventually the original low level of activity incorporated in plankton had become concentrated a million times when the eggs of water-fowl feeding on the plankton were analysed.

Incidents of this type are by no means restricted to the U.S.A. The level of Iodine-131 discharged by the Windscale reactor has to be very carefully monitored because of its incorporation by the seaweed growing in the area. Under normal circumstances this would not matter greatly, as the half-life of this isotope is only eight days and any radiation has dropped to one-eightieth within two months. Any possible hazard is greatly reduced as the effluent is piped 2 miles out to sea. In this particular area, however, the seaweed is collected and taken to South Wales for making into the local delicacy 'laver bread'. The original levels of iodine discharged were based upon the possible amounts of 'laver bread' likely to be eaten. The figures had to be reduced when it was found that some people were eating more 'laver bread' than had originally been anticipated.

Observation of the regulations concerning discharge of waste in this way is therefore dependent on considerable knowledge of the local tidal flow and currents, combined with information on the life-cycles and metabolic processes of animal and plant life. Adverse effects are unlikely if only isotopes with very short half-lives are dealt with in this way, but substantial dangers exist when half-lives are of more than a few days.

The dangers from radiation are due to its ability to disrupt the normal functioning of the cells in the body. Reproductive cells are particularly sensitive and, if subjected to even small doses of radiation, can lead to deformities in offspring which can be perpetuated through many generations. Large doses of radiation can cause death due to the breakdown of such a large proportion of the cells in the body that it is incapable of carrying out normal functions. As a result of studies carried out on the survivors of the atomic explosions on Hiroshima and Nagasaki, as well as on the victims of several nuclear accidents, a fairly considerable amount of information exists concerning the response of human beings to atomic radiation. It all shows that every effort should be taken to avoid being subjected to increased radiation at all times. As mentioned previously, there is no 'safe' level for atomic radiation.

It has been calculated by the U.S. Atomic Energy Commission that the probability of a *hazardous* accident occurring in a nuclear power reactor is less than 1 in 100,000 per reactor per annum. If we take this figure at its face value, in the light of known failures and dangers, it is not as generous as it appears. The figure was calculated on the basis of 100,000–200,000-kilowatt power stations whilst current concepts are in terms of five times this scale. Throughout the world about two hundred nuclear reactors are in operation, under construction or planned. So by 1980 the chance of any one of these suffering a hazardous accident is 1 in 500 in any one year. During the lifetime of these two hundred reactors (some of which have already been in operation for more than ten years) there is a 1 in 20 chance of a major accident occurring somewhere in the world. The odds are reduced below this if further reactors are constructed. From this it can be seen that it is wrong to consider that atomic reactors are the ultimate in safety. Nations must decide whether they are prepared to accept this relatively low safety margin to obtain the rather doubtful benefits from nuclear power.

Amongst the very useful side-products from nuclear reactors are radioactive isotopes. There is no doubt that the use of these in hospitals and scientific research has played a major part in prolonging the lives of many people as well as helping to raise the standard of living by indirect means. Certainly there is still considerable scope and justification for these uses. In a similar way it would be fair to say that, so far, nuclear power stations have shown themselves to be more of an asset than a liability, although the original hope that they would provide a cheaper source of electricity has long since vanished. Also it is questionable whether it is possible to justify further construction of such power stations if waste disposal problems are going to increase at a time when international controls and restrictions are becoming progressively tighter.

Some authorities consider that air and water pollution from conventional sources are of far greater danger than the so called 'hypothetical' dangers that might arise from nuclear power stations. This is a head-in-the-sand attitude, as such dangers are very real and have become reality on a limited scale in a number of places throughout the world. So far, safety precautions, and good fortune, have prevented a major catastrophe from occurring and creating a nuclear holocaust. The realities of the position are such that there is always a danger that a nuclear power station might suffer from an earthquake or have an airplane crash on to it, as can occur with any other building. So far the general public has been prepared to accept this 'hypothetical danger', but there will undoubtedly be a major outcry if it becomes a reality in the future. It should be appreciated that the cost of compensation following a nuclear disaster will be far greater than that for a conventional accident. In the case of the fallout which fell on the *Lucky Dragon*, killing one fisherman and hospitalising twenty-two others, the U.S. Government paid out $2 million in compensation. This is minute compared to the cost involved if, say, the nuclear-powered merchant ship *Savannah* blew up while in port. We can only be thankful that, so far, no nation has been foolish enough to develop a nuclear aircraft to increase the hazards.

Many conservationists support nuclear power stations on the grounds that they are smaller and cleaner than conventional power stations. This attitude ignores the fact that the former are usually built in areas of outstanding beauty and will remain as radioactive white elephants after their relatively short life-span. Conventional power stations have the advantage that they can be sited near to large cities, cutting costs and obviating many unsightly pylons. Modern techniques can in any case eliminate most of the air pollution traditionally associated with electricity generation.

In 1968 nuclear power supplied 12 per cent of Great Britain's electricity requirements, and only 1 per cent of the electricity needed in the U.S.A. Even at that level the problems of siting the power stations, disposal of wastes and cooling of water are gigantic. The answer to the need for more electricity is not necessarily more nuclear power stations, but rather new methods of generating it. The French have already shown what can be done by using tidal flow in their Rance River project opened in 1966. Long-term conditioning of our minds in favour of nuclear power has made us almost totally ignore the possible development of such methods as tidal flow or thermo-electric generation. Current plans for increasing nuclear power generation will result in a corresponding increase in radioactive waste.

Our legacy for the future already exists in the 250 square

miles of Nevada that will remain unusable for the next 24,000 years due to contamination from plutonium released during atomic tests. To this we intend to add further radioactivity in the sea and the atmosphere. There is no safe level for radioactivity, and we are progressively reducing the margin between the region of no obvious injury and that of observable changes in blood cells. Whilst any increase in the unavoidable background radiation is very small it is totally undesirable. If we are to continue to use nuclear power we must be honest enough to admit that there are dangers just as there are in driving a car or smoking cigarettes.

10 The thunder of civilisation

If radiation is the most insidious destroyer of the environment then noise is the most blatant. People everywhere are growing conscious of the substantial increase in noise of all kinds that is being imposed upon them with apparently little or no effort being made to ameliorate the situation. As a nation becomes more mechanised in its factories; and its motor transport increases, along with the advent of supersonic aircraft, so noise levels build up. The overall result is that everyone is expected to tolerate more noise for longer periods of the day.

Unfortunately it is not possible to assume that a noise that is objectionable to one person is necessarily also objectionable to another. And what may be excessive noise at one time of day may be more acceptable a few hours later. Unlike water pollution and air pollution, with which it is possible to say that certain levels of pollutants are undesirable, the objection to noise is frequently subjective and not necessarily related to a scientific measurement. Whilst all noise is sound, not all sound is noise. Sound only becomes noise when the listener finds it objectionable. So that whilst Beethoven's Fifth Symphony may be superb music to one person it may be considered a hideous row by somebody else who prefers pop music, or vice versa.

Measurement of sound can be carried out in three different ways: by the study of its pitch (frequency as cycles per second), its volume or intensity (expressed in decibels), and its duration (measured in conventional units of time).

It has become common practice in the newspapers for noise levels to be referred to as being so many decibels. This is an oversimplification of the true state of affairs, as the method of calculation of decibels requires the choice of some arbitrary zero level from which all measurements are taken. Fortunately confusion is avoided, as the base figure is taken at the normal lower level of human hearing, although sounds below this level are detectable.*

As is seen from the footnote, the decibel scale for measurement is not simply an additive function. That is, an increase from 100 dB to 120 dB does not represent an increase of 20 per cent but an increase to ten times the original level. This is because the decibel form of expressing value is not a linear scale but a logarithmic one. This fact has tended to obscure the true extent to which noise levels have increased. An increase from, say, 120 dB to 126 dB does not look too bad until it is appreciated that this actually represents a doubling of the noise level. In fact the noise level doubles every time there is a 6 dB increase. The true relationship between increase in noise levels is shown in the following table:

Increase measured in dB	True increase in noise level
20	10 times
40	100 times
60	1,000 times
80	10,000 times
100	100,000 times

Noise levels below 40 dB are very low indeed as normal conversation has a noise level between 55 and 60 dB, whilst a shouting voice will attain about 80 dB. This means that objec-

* The lower limit of detectability of sound for the average human ear corresponds to a pressure of 2×10^{-5} Newton/sq. metre. Noises are related to this according to their sound pressures by the equation:

$$\text{Sound Pressure Level} = 20 \log \frac{p}{p_0} \text{ decibels.}$$

tionable noise levels can be considered to occur roughly above 70 dB. Besides the decibel other units for sound measurement, such as the 'phon' and the 'sone', have been introduced, but the use of these is normally restricted to technical literature dealing with acoustics. They can in any case be related to the decibel, which is used in most general articles on the subject.

The detection and transmission of sound through our ears is a very involved process. Ears are very sensitive and delicate instruments which receive vibrations travelling through the air and transmit them by nerve impulses to the brain. There are three sections to the ear, each vital to the hearing process. Sounds pass through the external ear (which forms the ear trumpet) and arrive at the ear drum. From here the vibrations are passed to the middle ear, which consists of three small bones, the malleus, the incus and the stapes. These in turn pass the vibrations to the inner ear of which the most important part is a small coiled tube called the cochlea. The cochlea tapers along its length and terminates in a collection of nerve fibres which form the auditory nerve to the brain.

Damage to the ear can, therefore, occur from the deterioration or destruction of any of these composite parts. Apart from accident and infection, which can both lead to loss of hearing, noise itself is a major factor in the development of deafness. Subjection to high levels of noise for short periods of time can lead to temporary deafness, recovery from which may take several days. Permanent deafness can occur if part of the ear is ruptured. In addition, continuous exposure to high noise levels leads to the ears becoming less sensitive to sounds, so that effectively all noises are heard as if they are much softer. This form of deafness, frequently caused by noisy industrial workrooms, develops from damage caused by the noise to the cochlea.

It has been found that ear injury can result from short-term exposures to sounds above 105 dB. Subjection to sounds above the 120-dB level automatically create a painful sensation in the ear. Levels of 150 dB are considered to be a general danger to health leading to hypertension and heart troubles in addition to ear damage.

But it is not only the volume of noise which can be objectionable, for people can equally be irritated by a continuous high-pitched hum or regular banging sound. This means that the frequency (i.e. the pitch) of the noise can be just as important as the volume. A noise-measuring instrument which only responds to one particular frequency will obviously give no reading if the noise signal is of some other frequency which it has not been constructed to record. For this reason most modern sound-level meters are designed to read the contributions of several different frequencies to the noise level simultaneously, in order to give a reading corresponding to that actually heard by a human ear. For this purpose four different frequency weightings are used, designated as 'A', 'B', 'C' and 'D' scales. The 'A' frequency weighting is employed in most general-purpose instruments and a noise measured on one of these is referred to as having a dB(A) value. The 'D' scale has recently been introduced for particular application in measuring aircraft noise.

Noise-measuring instruments using microphones and amplifiers were first developed over forty years ago. Essentially the same arrangement of microphone and amplifier with a recorder is still used today, although modern electronics with transistors and printed circuits have enabled smaller portable instruments to be constructed. This fact alone will enable enforcement of noise control to be more readily carried out in the future than it has in the past. It also means that the cost of such instruments is well within the means of local authorities and private bodies concerned with establishing the value of local noise levels causing a nuisance.

It should be emphasised that different instruments are used for the various kinds of noise that can occur. For instance, a jet aircraft produces a continuous noise covering a very wide band of frequencies, whilst a metal body-pressing machine produces a short-term impulse noise and road traffic gives a variable intermittent noise. By using the wrong measuring device it is possible to obtain totally misleading indications of the noise level, despite attempts to overcome this in general-purpose instruments.

In Great Britain noise control and noise abatement are governed by Common Law, local bye-laws and the Noise

As demand for land increases, reclamation of derelict areas is becoming more important. These photographs show what has been achieved on an 11-acre site of old colliery spoil heaps at Aspull, Lancashire, giving a tremendous improvement in appearance as well as providing much needed building land.

With the increase in the number of flights and the size of aircraft people who live near major airports are subjected to a continuous barrage of noise every day.

The other type of jumbo – in fact Bangkok elephants kept at the Windsor Safari Park Zoo – found the noise from London (Heathrow) so intolerable that they stampeded. As a result they had to be fitted with earmuffs.

All too often the alternative to aircraft at the bottom of the garden is an elevated motorway at bedroom level. London's Westway cuts a swathe through heavily populated residential areas, subjecting thousands of people to the incessant noise of road traffic.

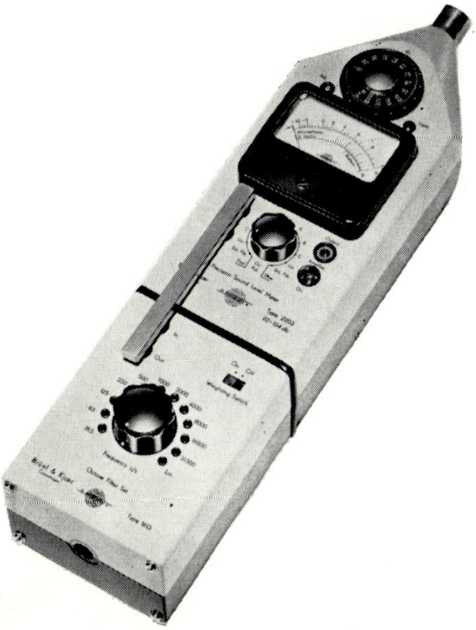

Enforcement of laws limiting noise depends upon accurate means of measuring noise levels and intensities. The improved designs in portable precision sound level meters of the type shown here will play an increasing role in this realm.

Nuclear power stations at present take up large areas of land on remote coastal sites or in National Parks. Increasing demands for electricity throughout the world are causing many nations to consider the siting of future nuclear power stations nearer to centres of population.

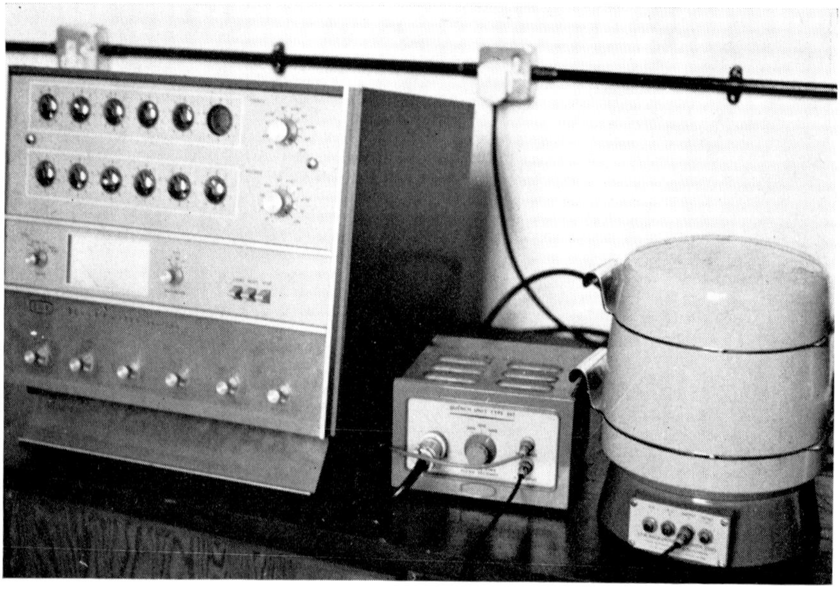

Measurement of radioactive materials from atomic reactors or natural sources is carried out under carefully controlled conditions. Samples enclosed in the lead chamber can be left to emit their dangerous rays, which are measured and automatically recorded by rotating lights on the counting instrument.

Abatement Act, 1960, as well as by the Public Health Act, 1936. Some control, in theory, exists under the Motor Vehicle (Construction and Use) Regulations, 1950, but this would appear to be almost totally ineffective. Noise standards and controls operated throughout the world vary greatly, but attempts are being made to achieve consistency through the International Organisation for Standardisation. Already several European nations are employing a single established procedure for checking noise from motor vehicles.

Under Common Law any person can apply for an injunction against any individual, organisation or company if he feels that noise being created interferes with his comfort in his own home. Decisions made in such cases have to take into consideration a variety of factors including the general industrial or residential character of the neighbourhood as well as the time of day when the nuisance is created. In a recent case in London a woman was unsuccessful in trying to curb the night-time noise from a building site as this was the only time that the work, above an underground railway, could be carried out.

A more successful action was taken in February 1970 by a body of school managers and the Hampshire County Council against noise from a car-testing circuit at Andover. In this case, which reached the High Court, the judge said there was evidence that lessons at the school had to be halted because of the noise. As a result a ban on testing between 9 a.m. and 3.45 p.m. during term time was imposed against three of the four defendants.

Local bye-laws against noise nuisance have been introduced throughout Great Britain since the Municipal Corporations Act, 1882, gave this power to deal with noisy animals and street hawkers. An increase in the number of such bye-laws came about as a direct result of the publication of the *Committee on the Problem of Noise: Final Report* which sat under the chairmanship of Sir Alan Wilson from 1959 to 1963. This suggested model bye-laws which local authorities could adopt to deal with the, then, growing problem of transistor radios, record players and live music performances. Generally these recommendations suggested that restrictions should be imposed between the hours of 11 p.m. and 6 a.m. Some local authorities adopted the general recommendations with variations in the hours of restriction.

Until 1960 the main protection against noise nuisance was under Section 99 of the Public Health Act, 1936, by which any three householders can lodge notice of complaint in cases in which the noise is likely to be dangerous to health and is capable of being prevented or controlled. The Noise Abatement Act, 1960, was a major step forward in reducing noise levels in that it tied up many of the loose ends and established some degree of uniformity throughout Great Britain. Its weaknesses are that it imposes no restrictions on aircraft noise or upon the statutory undertakings that are in so many ways a law unto themselves. This means that complaints against such bodies have to be made under one of the earlier Acts. Another weakness of the 1960 Act is that on conviction the maximum penalty is a fine of £200. This is negligible for a building contractor or manufacturer carrying out night-work in order to complete a contract on time and undermines the purpose of the law.

It is certainly far from easy to obtain redress against the creation and continuation of noise. The difficulties of proving that a nuisance has been caused; of getting neighbours to support an action; or of proving that a business is not using the best possible means to control the noise are enough to daunt the most aggrieved person. Nevertheless the increased agitation against noise levels has resulted in more research into the origins and abatement of noise and may eventually lead to further legislation. Unfortunately, as in the field of water and air pollution, the advocates of noise control will find themselves up against the people who believe that noise is synonymous with industry and production. In fact continuous industrial noise at high levels can lead to a lower level of efficiency, the occurrence of more mistakes and a greater turnover of workers, quite apart from purely humanitarian considerations.

No legislation exists to protect workers against noise, and the Wilson Committee felt that legislation on this was unnecessary. The worker, whether in factory, office or shop must, therefore, resort to Common Law or union action in an effort to obtain reasonable working conditions. Despite this deficiency radical improvements in noise abatement have been brought about in factories, often with startling effects upon productivity and morale. It is often forgotten that a person at work in a factory is not only subject to the noise from his own machine or labours, but also from those around him and from other extraneous noises arising from ventilation fans, vehicles, conveyor belts and telephone conversations. So that whilst his own noise creation might by itself be perfectly tolerable the cumulative effect of all the noises may be dangerous.

Noise control can be carried out anywhere between the source of the noise and the receiver. This means there are three possible areas for action; the source itself can be controlled, the absorbance of the transmission path between source and receiver can be increased and the receiver itself can be insulated. All these methods will need to be widely employed if tolerable noise levels are to be obtained in industry. In his annual report for 1969, the Chief Inspector of Factories claimed that few industries were without noise problems. In many cases these can be overcome by improved machinery, modern processes and better workshop layout.

Source control involves either reduction of the initial noise level by improvement in the design of the mechanical device, or by absorbing the sound as it is created. Just a little thought given to design can bring about improvements. The replacement of a metal unit by plastic or rubber can eliminate metal-to-metal impacts. Use of workshop vehicles with rubber rims instead of unyielding iron have been shown to lead to major reductions in noise levels. Similarly, erection of machines on soft flexible mountings can dampen vibrations and prevent transmissions throughout a building. In some cases the change from a purely mechanical machine to a more automated one is often the simplest way of bringing about improvements by source control. This is particularly obvious in open-plan offices in which electric typewriters have not only made working conditions much more amenable but have led to an improvement in the quality of typing and an increase in output. Similarly the construction of extractor fans with many parts made from moulded plastics have meant reduced extraneous noise interfering with conversation.

Source control of noise can take rather unexpected forms, as has occurred in the case of car tyres. Research on tyres showed that a great deal of the noise that tyres produce on some types of road surface arises from resonance occurring due to the regular design of the tread pattern. By introducing an irregular pattern substantial reductions in noise were achieved and new tyres are now being manufactured thus modified.

Source control is obviously the only way in which noise from a moving vehicle, or any other mobile object, can be reduced or eliminated for everyone's benefit. Car and motor-cycle exhausts come in this category and are high on the list of sources of annoyance. The purpose of the exhaust pipe is to conduct the waste gases away from the engine. In so doing it also conducts the noise from the engine and amplifies it. Because of this exhaust pipes are supposed to be fitted with silencers in the form of expansion boxes in which the pressure of the waste gases is reduced and the potential noise level lowered. Whilst the majority of manufacturers attempt to maintain a low exhaust noise level, the Motor Vehicles (Construction and Use) Regulations, 1950, is inadequate for effective control to be carried out. Disregard for these regulations is shown by the twelve thousand prosecutions made during 1969. The Wilson Committee recommended that maximum noise levels for moving vehicles should be 85 dB(A) for motor cars, and 90 dB(A) for motor cycles. No action was taken by the Ministry of Transport on this until the end of 1969, despite the fact that noise from motor vehicles is one of the major annoyances suffered by people today. Apart from roadside checks on noise levels, this problem could be overcome by adding an exhaust noise test to the annual vehicle test that is carried out for a roadworthiness certificate.

Another source of noise in cars is the slamming of car doors. This is entirely a design fault as slamming is frequently necessary, even in modern cars, to ensure that the door is correctly shut. Some manufacturers have eliminated this by using door locks which are self-tightening, and will ensure a secure hold on the door when pulled gently to a closed position.

Interior car noise is a contributory factor to driving fatigue and it is surprising that it has received relatively little attention from the motoring public. Unfortunately few motor cars are manufactured to Rolls-Royce standards, with only the sound of the clock ticking. Some interest has been sparked off by the Automobile Association including interior car noise amongst the large number of features studied in car reports for *Drive* magazine. For the majority of mass-produced vehicles the noise level exceeds 68 dB at 30 m.p.h, and is frequently above 80 dB at 70 m.p.h. Most of us subject ourselves to more noise than is necessary in our cars, and there are soundproofing kits available for most models. The soundproofing of a car involves lining the bonnet and floor of the vehicle with fibre padding to prevent the transfer of noise throughout the vehicle.

Very little appears to have been done to eliminate railway noise and people in many parts of the country frequently have their sleep interrupted by the banging and clanking that is permitted in shunting yards. Most of this noise does not come from the engines doing the shunting but from the couplings between the trucks and carriages. Railways could make substantial contributions to the quietness of the night by having buffers and linkages coated with a substantial layer of durable plastic. But then British Rail is virtually a law unto itself and public opinion so far lacks legislation to back it up in this matter.

The same position exists over aircraft noise. It is quite incredible to hear supposedly responsible people say that people must learn to live with the noise of jets and supersonic transport. One author has implied that people must expect noise if they have purchased houses built near airports – he forgot to mention that when many of the houses were first built the noise levels were tolerable, but that the introduction of jets in the past twenty years has led to these areas becoming almost uninhabitable. It is about time that we realised that only a minority of people has ever travelled by aeroplane, and the percentage of regular travellers is minute. So many thousands of people are having to tolerate unreasonable noise levels for the sake of a minority who want to save an hour or so for hurried business trips or extra sunbathing. There must be something wrong with the world's priorities for this state of affairs to have been created. Noise levels are increasing from aircraft, not because it is completely unavoidable, but because when engines are developed research is directed to making them even more powerful instead of trying to reduce the noise levels on engines already giving perfectly satisfactory service in established aircraft. According to the Wilson Report the bulk of expenditure for research on reduction of aircraft noise, about half a million pounds per annum, comes from the Government. The aircraft industry itself only spends between a quarter and one-third of this amount. Studies carried out at London's main airport, Heathrow, showed that between 20,000 and 40,000 people lived within the 50 NNI contour.* The Wilson Committee said 'We have agreed that the noise to which many people near London (Heathrow) Airport are subjected is more than they can reasonably be expected to tolerate'.

The present measures aimed at noise reduction require a steep climb at take-off and an early reduction in power. Whilst this is not intended to endanger the aeroplane, it does decrease the safety margin, and from that point of view would be better avoided if possible. Whilst this may help people near the airfield to obtain more sleep, it does not really deal with the problem. There is an urgent need for greater efforts to reduce aero-engine noise. When the incentive is powerful enough technology and science have generally been able to overcome even the most insuperable problems. It would appear that a little more incentive in the form of Government pressure is required.

One substantial step forward was the introduction of noise certificates for aircraft by the Board of Trade in May 1970. Great Britain was the first nation to implement this, after recommendation at an international conference in 1966, and will almost certainly be followed by other major aeronautical nations. The certificate permitting operation will only be

* 'Noise and Number Index' (NNI) figures relate the noise level of aircraft at a place with the number of landings and take-offs passing over that point. The Board of Trade reported that the average person considered annoyance 'moderate' at 44 NNI and 'considerable' at 60 NNI.

granted to new conventional jet aircraft which meet standards for noise levels that are 50 per cent less than for contemporary aircraft currently in use. This will come into operation in January 1972.

At the same time it cannot be denied that some efforts have been made to make life a little better for local residents around airports. Stationary jets under test are screened by specially built earth banks and engine mufflers, whilst flight paths are arranged to take aircraft along the least-populated routes. It was concern over the noise level at London (Heathrow) that led the British Government to establish soundproofing grants to residents within the 55 NNI contour. These grants were, however, limited to a maximum of £150 and applications had to be made before 31 December 1972 in order to be considered. In the meantime greater international concern over aircraft noise is likely to lead to noise restrictions becoming more common in the relatively near future. This possibility is of great concern to the manufacturers of supersonic airliners.

The true potential of the supersonic boom is only just beginning to be appreciated by the general public now that prototypes are undergoing their test flights. The Celtic fringes of Great Britain and Ireland are being subjected to periodic doses of eardrum bombardment just to prove to them that they cannot escape civilisation by living in the few remaining unspoilt areas of the country. The noise of the boom can actually be detected within a path of 25 miles wide on either side of the path of the aircraft. It creates a series of high-pressure sound waves which in certain cases can exceed normal atmospheric pressure by 3 pounds per square inch. This increase in pressure occurs very rapidly and is considered to be capable of causing severe property damage by some experts. At present evidence on this is conflicting, and many conservation and environment groups are busy collecting information to gain a true indication of the possible danger from supersonic flights.

It may well be that supersonic transport is here to stay. But it should be firmly stated right from the beginning that *all* supersonic booms should be carried out at least 100 miles away from land. The final stages of journeys to all continents should always be carried out at subsonic speeds. International agreement on this could be an important factor in reducing aircraft noise, as it has been claimed that when operating below the speed of sound the Concorde's engines will create less noise than those of conventional jet aircraft.

The second method of noise abatement, by absorbing the noise during its transmission, is particularly applicable to manufacturing equipment constructed with an intended life-time of twenty or thirty years. Thus without some means of noise reduction any undesirable sounds created by the machinery will be with us for years to come. In these cases the use of suitable baffles and absorbers or deflectors can often overcome the difficulties and lower the immediate noise level around a machine operator by as much as 20 dB. A reduction of 10 dB alone corresponds to a lowering to one-third of the original level and can often be achieved by even a cheaply constructed baffle.

The abilities of building materials to absorb sound depend upon the nature and thickness of the construction. For technical purposes the absorption characteristics of materials are expressed by the term 'absorption coefficient'. This is the ratio of the energy absorbed by the surface to the energy incident on the surface. Values for absorption coefficients range from nearly zero for bad absorbers to the highest possible value of 1·0 for the best absorbers such as fibreglass. The value of the absorption coefficient usually varies with the frequency of the noise, and it is found that whilst some substances will absorb low-pitched noises easily, the high-pitched noises are not so well absorbed. Most glass comes in this category, and a window made from glass one quarter of an inch thick will lead to a reduction overall of about 20 dB but will absorb more of the low sounds whilst still transmitting the majority of the high sounds. Double-glazing of windows can lower the noise level by 40–50 dB, depending upon the air space between the two panes and it is still the high sounds that tend to be transmitted.

High-quality sound-insulating materials find a wide variety of uses. That noise can be greatly reduced has been shown in the design of airliners in which passengers travel in considerable comfort through being effectively shielded from the hideous row of the engines only a few feet away. Sound-insulation of this type is achieved by such features as double-glazed windows of special glass with the interior pane mounted on rubber supports. In addition to this the main passenger cabins of aircraft have floating walls carried on vibration-isolation mounts. These help to prevent external vibrations from being transmitted through luggage racks and seats. From this it will be appreciated that the means are often already available to enable substantial noise reductions to be achieved when the wish to do so exists.

Effective screening by trees and cuttings has been one successful approach to dealing with motorway noise. Where the new road replaces an old one along the same route a reduction in noise level often results. This is because the traffic on the new road moves continuously as a result of the flyovers whereas before it was stopping and starting because of road junctions and traffic lights. This was reported to be the case observed by some of the occupiers along the route of the southern approach to Blackwall Tunnel under the Thames. At the same time the construction of an elevated motorway along a new route leads to a considerable noise level being created where very little existed before. This was certainly the case in Acklam Road, North Kensington, where the Western Avenue extension runs only a few feet away from the bedroom windows of local residents. In this type of situation there seems to be some need to build up an absorbent baffle wall along the motorway, or to ensure that houses within noise range are either demolished or fully soundproofed. It has been reported that noise levels attain nearly 80 dB in properties close to this particular motorway and that the daytime level is continually well above the 50-dB level considered intolerable by the Wilson Committee.

Acceptable levels for noise that the Wilson Committee suggested for different areas were:

	Day	Night
Country areas	40 dB(A)	30 dB(A)
Suburban	45 dB(A)	35 dB(A)
Busy urban	50 dB(A)	35 dB(A)

It also said that these should not be exceeded for more than 10 per cent of the time. There are still many districts in which people are suffering from noise far in excess of these levels.

Absorption of noise by baffles and acoustic materials features prominently in many modern buildings. Boardrooms are fitted with acoustic tiling to help keep conversation noise levels down and to prevent information from being overheard. Concert halls and theatres are specially treated to prevent reflections occurring from hard surfaces and causing echoes during performances. Such methods are often necessary in older buildings which were constructed before knowledge of acoustics was as extensive as it now is. This is the case in the Albert Hall, in which the domed circular auditorium would be unsuitable for concert performances without the extensive acoustic modifications that have been carried out. A very common example of eliminating undesired extraneous noise is in the telephone hoods which are to be found in open places such as works canteens and factory production areas. Prior to their introduction telephone conversations carried out in a room packed with several hundred people eating and talking or with machinery humming away were almost unintelligible.

The third approach to sound control is to absorb the sound at the point of the receiver. This is often the last resort for people who are unable to achieve effective control at source of some objectionable noise. The simplest form of device used is earmuffs; very necessary for airport workers who often have to be in close proximity to aircraft engines during testing and warming-up periods. Earmuffs are all too often necessary for people living on main roads in order to obtain a reasonable night's sleep. Their use will increase if proposals to increase the permitted gross weight of lorries are implemented, though even they can do nothing about the ground vibrations that these juggernauts create.

Effective sound-insulation is of considerable importance in terraced housing and multi-storey blocks of flats. The need has led to the Royal Institute of British Architects adopting a British Standard Code of Practice to be applied to architect-designed dwellings. Close community living requires special attention to be given to sound conduction along reinforced concrete floors and to the penetration of conversation through walls of adjacent dwellings. In many cases the correct solution lies in the choice of the construction materials employed as well as the nature of the building. It is, for instance, well known that one of the best sound insulators is a cavity brick wall, whilst a plywood partition is virtually ineffective.

Most sound from external sources enters a house through the windows and substantial noise reduction can be achieved if these receive some attention. Double-glazing of windows can lead to a lowering of 30–40 dB if an air space of several inches is held between the two sheets of glass and the glass is mounted on a sound-absorbent sealer. For soundproofing of windows to be fully effective there must be no open windows for letting in fresh air, so air conditioning is also necessary. The whole matter of soundproofing houses has now become of such great importance that the publishers of some home-improvement magazines have devoted large sections of their publications to methods available for soundproofing walls, windows and roofs as well as for building the necessary soundproof ventilators.

Although Government grants are now available for people living near London (Heathrow) Airport the maximum amount of £150 is inadequate for properly soundproofing more than one room, so a considerable incentive exists for householders to tackle the job for themselves. The grants for this purpose were introduced by the Labour Government of 1964–70. It is time this was reviewed and possibly extended to cover up to 50 per cent or £300 (whichever is the smaller) for soundproofing any house in which either daytime or night-time noise levels substantially exceed the Wilson Committee recommendations. It is totally unreasonable for any government to allow citizens to be subjected to increasing amounts of noise from supersonic aircraft, mammoth lorries, diesel locomotives and motorways without providing help to restore decent living conditions. It is just as possible to designate 'black areas' for noise as it was for smoke pollution and to deal with them in a similar manner.

A growing movement against the placid acceptance of more and more noise is obviously developing throughout the world. Already in the U.S.A. six hundred householders have won a major court case for compensation from the City of Los Angeles for having deprived them of the free use of the air-space above their homes. As a result the city was directed to pay a total of £300,000 compensation based upon the deterioration in house values and the difficulties caused by mental strain. A similar case in Great Britain is probably not possible because of the rights given to airlines under the Civil Aviation Act passed nearly fifty years ago before current noise levels were envisaged.

If the silent majority wish to continue to enjoy their silence then for a little while they must become more vocal. Noise, like every other form of pollution, is a health danger – it can create deafness and nervous disorders, it can wreck family life and ruin the future for children unable to study. Noise abatement is not something that just concerns a few cranks. It concerns all except the totally deaf, whose numbers we are likely to join if we continue to ignore the pounding to which our ears are subjected.

11 Narrow is the path

It is quite obvious from the proliferation of articles and books on pollution and the environment that it is very easy to adopt a completely pessimistic attitude. All around us are prophets of doom, making our flesh creep and trying to out-think each other in devising the worst form of slow death to cover the world. This might initially have some value to it if it makes people more aware of the potential dangers around. It could, however, be carried too far and create a panic situation in which decisions are taken for short-term expediency with inadequate appraisal or lack of consideration for long-term objectives.

Already, on a national scale, many hundreds of scientists are involved with investigating and devising methods for combatting pollution and improving the quality of living. Throughout the world similar work is being carried out in all the industrialised countries, involving millions of man-hours and vast sums of money. The problems of pollution were not created overnight, nor can the damage that has been done be repaired or reversed by a newspaper article or even an Act of Parliament.

Apart from explaining the nature of the problems that beset us, this book has been concerned with showing what has already been carried out to improve the situation and with indicating where there is scope for further action. The worst possible thing that could happen is for the momentum engendered by European Conservation Year to fail to be carried forward into the future and for public concern over the environment to drop. This is unlikely to be the case so long as there is a desire for clean beaches, improved leisure facilities and for less noise.

Only public opinion can lead to further improvements being carried out. Local authorities should not be able to put the construction of a new city hall before the building of a sewage works, or a theatre before a refuse disposal plant. But this type of decision will continue to be made while local prestige is considered to be more important than local health. Further procrastination will lead to progressive deterioration in the quality of living.

There is already evidence from the U.S.A. that the continual rise in the expectancy of life has now stopped and appears to be on the decline. Whilst the causes are still uncertain, the decline has been attributed to the combined effect of the increase in pollution, the presence of dangerous chemicals in air, food and water, and the stress of modern living. It indicates that we may have reached the threshold of tolerance for the human body to cope adequately with a combination of contaminants. Whilst the natural protective and rejection abilities of the body can often deal with fairly large quantities of dangerous substances, there is very little information on the effects that mixtures of small amounts of many different poisons will have.

Also we cannot afford to just stand still and accept the present number of dead rivers and derelict sites as reasonable. Increasing populations will automatically lead to increasing quantities of sewage, more cars and greater demands for homes and raw materials. Standards for pollution control will have to get very much tighter. Failure to grapple with the problems will mean fewer rivers for boating and fishing, poorer quality drinking water, more contaminated beaches, increase in respiratory disorders and more industrial land dereliction. Not only will we have a '*silent spring*', but a joyless summer, a smoke-bound autumn and a desolate winter into the bargain.

It is obvious that greater international agreement is necessary on many aspects of pollution if effective control is to be obtained. This is particularly so in countries which share the use, and misuse, of rivers, and over the control of navigable and international waters. Future decisions here will depend upon the deliberations of the large number of committees sitting under the auspices of the United Nations and the various European and professional bodies. It does appear that regulations on shipping and discharging of oil will be considerably tightened and that international agreement may lead to nations being prevented from discharging waste into tidal waters when this is likely to lead to contamination of a nation's seashore. Similar agreements concerning inland waterways are likely to cover rivers such as the Rhine. Success of such actions will depend greatly upon individual countries maintaining a high level of control and inspection.

Individual nations have the power to do much more for themselves in a shorter period of time and probably with more effect, as their decisions do not have to be taken in the light of international compromise. This is not entirely the case in the U.S.A., where the fifty individual states have powers to introduce their own state laws which may conflict with Federal Government intentions. On rare occasions state legislation may be in advance of Federal law, as with California's controls on motor-vehicle emissions. But this only emphasises the difficulties of uniformity where a number of legislative bodies are involved. In Great Britain this situation does not exist and is probably one of the reasons that the pollution problem has not attained the magnitude that it has in the U.S.A. At the same time the much smaller size of Britain means that the effects of pollution are less easily hidden from public view.

It is quite possible for further improvements in the British environment to be achieved without resorting to much additional legislation. This is because a great deal of the power required already exists in the hands of local authorities and government bodies. Differences in conditions which prevail between various areas of the country are due to some authorities having used their powers to greater extents than others. Considerable pressure will be necessary to ensure that further smoke-control zones are established rapidly enough, not only to achieve elimination of the Black areas mentioned in the Beaver Committee interim report, but also to re-establish zones suspended because of the shortage of smokeless fuel. At the same time this will create considerable antagonism from local residents if control of industrial pollution fails to keep in step with regulations on private householders. The Alkali Inspectors have a major part to play in this and should take a more careful look at atmospheric fallout which in the past has been disregarded as being 'psychological'!

A similar tightening up on the freedom of the 'statutory bodies' is also needed. The nation cannot continue to permit their large-scale desecration of the countryside with nuclear power stations in national parks, gigantic gasholders in the middle of small towns, enormous electricity pylons striding across the open countryside, and beautiful valleys being flooded for reservoirs without having a much greater say in where such things will be situated. This means that there should be a greater obligation on these bodies to ensure that details of planning proposals are released to all local authorities as early as possible, and certainly before any commitment to capital expenditure has been taken. The multiplicity of laws covering the statutory bodies require simplifying and planning responsibility needs to be consolidated into one Act of Parliament.

Action is also needed to prevent the discharge of untreated sewage into rivers or close to beaches. A national programme should be established covering from five to ten years, during which time every local authority should be obliged to construct treatment plant capable of dealing with the anticipated needs for twenty or thirty years into the future. This should be co-ordinated with similar controls on industrial water pollution, which should also be progressively phased out along the length of rivers, starting from their sources and working downstream.

Further industrial waste heaps should only be permitted on a limited life basis on the understanding that each heap is removed at the end of the time granted. There should also be a Government body responsible for finding methods of disposing of disused waste heaps in as economic a manner as possible with the co-operation of the company or authority responsible for creating the heap in the first place. Such a body will need to have certain powers of compulsion for situations in which co-operation cannot be obtained. Authority to deal with disused excavations could also come under this, but these are much less of a problem because of their value for boating lakes and fishing. Coupled with this is the need for improved methods of waste disposal. This again will require a national programme to ensure that modern plant and equipment is available in all areas to deal, not only with household refuse, but with garden waste, old furniture, shop waste and some industrial waste. The G.L.C. has already demonstrated how one purpose-built plant can be used to carry out the work originally done by a large number of small, limited-capacity incinerators. What will be required are more joint ventures between adjacent local authorities instead of small localised destructors.

Actually, throughout the whole realm of the subject of pollution, it is possible to see that a great deal more can be done; and a great deal has been left undone because of local jealousies and the desire for each council to have, or not have, its own sewage works, its own waste disposer, as well as its own swimming pool, civic centre, stadium and technical college. In these cases one can see a clear-cut justification for completely overhauling the local government system in Great Britain. Certainly our present hotchpotch of Parish, Rural, Urban, Borough and County Councils, for all their historical origins, are inadequate and frequently incapable of dealing with the complexities of modern living. The British Government's proposals for the reorganisation of local government deserve full consideration for these reasons alone.

During the coming years more specialist companies will develop to advise industry on how to control pollution and deal with its wastes. At present this function is mainly restricted to individual manufacturers pushing their own pollution-control productions, and to recommendations from government departments about a particular process or manufacturing technique. Demand for tighter controls on gas emissions and limitations on dumping will necessitate teams of specialists who can come into factories and deal completely with the interdependent problems of water, air, solid waste and noise at the same time. It will become an automatic thing for the most suitable equipment to be purchased and not that which has had the best high-pressure salesman selling it. Industry will be faced with fairly considerable problems in meeting the tighter restrictions which will eventually be imposed. This will probably require either Government grants to help with the purchase of purification plant, or special provision on tax allowances to avoid substantial price increases to cover the increased overheads of the companies involved. Without this help, reduction of air and water pollution will be held up for years. As it is, the improvement will have to be spread over a number of years if productivity and jobs are not to be endangered. It must be remembered that few people are prepared to be put out of work because there is too much smoke from a furnace or because untreated waste goes into the river. No government that created unemployment by imposing impossibly tight pollution control could ever be considered to have acted in the best interests of the nation.

In all this the part of the individual is paramount. Millions of people should all be prepared to be the eyes and ears of the nations ready to act against unsightly planning proposals, doing their utmost to regain and retain pleasant and green countries with clean rivers, clear air and unspoilt landscapes. Perhaps that is too much to expect from any nation that makes a tradition of throwing waste paper on the ground. It may be too much to expect a nation to change, even within ten or twenty years, when it has taken fifteen hundred years to adopt the central heating first introduced by the Romans. But responsible people do want to do something and do want the green and pleasant land we sing about so frequently and thoughtlessly on Sunday mornings – and forget the rest of the week.

You, the individual, can support local councillors who are in favour of smokeless zones and better sewage treatment. You can ask for speakers on pollution at the Women's Institutes, the Round Table, Rotary Club and local discussion groups. You can encourage your children to respect the countryside and to bring their litter home, and you can lend your support to the large number of national organisations, usually with local branches, that are concerned with these problems. A list of these will be found in the Appendix and they will all welcome your support. A national problem can only be solved by national support.

Throughout the chapters in this book there is one recurrent feature – population. Whether considered from a national or an international point of view, the interwoven questions of pollution, conservation and the environment all come back to the same starting point – people. More people mean more human wastes to dispose of, while at the same time requiring more clean water for drinking and washing. They require more jobs which will mean desecration of more of the earth's surface for raw materials, and elimination of more waste into the atmosphere, the rivers and on to the land. The circle becomes larger and larger and more unwieldy as the population increases so that it becomes a fight just to stand still to stop standards dropping even further, without actually being able to make the situation any better.

Internationally the combination of world poverty, lack of food, unemployment and increasing population is so horrifying that it is difficult to see the way ahead. Even from the parochial outlook of Great Britain we find that our fifty-three million people live in one of the most overcrowded groups of islands in the world and are only capable of growing about 50 per cent of our total food requirements. Because we are dependent upon manufacturing and exporting in order to obtain our extra food requirements, we have developed industrial processes at the expense of the quality of our national environment. This will continue to be the case as our population increases and the competition for land between agriculture on the one hand, and demand for houses and industry on the other, will leave even less for recreation.

At present we are like a motorist driving at 70 miles an hour through a fog on a motorway and ignoring the flashing warning signals. We cannot continue to positively encourage an increasing population in the way we have in the past. Irrespective of colour, class, creed or national origin we should not deliberately inflate our population by permitting unlimited residence for citizens from other countries, usually less populated than our own.

Whilst immigration may help to solve some temporary shortage in manpower for particular industries, in the long run it creates far more problems than it solves. It is certainly not the answer for difficulties in countries such as Great Britain, which are already overcrowded. We shall find it difficult enough to provide the necessities of life for any natural increase of population in the future, and can only aggravate the situation further unless permanent immigration from all other countries is stopped.

At the same time every reasonable effort must be made to achieve population control on a world-wide basis. Obviously we can only make those decisions which affect us as a nation, and are in no position to dictate what steps other countries should take. Within our own country we have already taken the first steps to stop the wholesale distribution of family allowances to people irrespective of income. Similar steps should be taken over maternity grants and maternity allowances. Nobody would seek to avoid giving help to families with genuine need, however many children there might be, but it is an entirely different thing to give all families a national bonus for providing more mouths to feed. We should be quite clear that unless the voluntary limitation of families is successful, not only will living standards drop and food requirements increase, but at some time in the near future it may lead some government to impose a compulsory control on the numbers of children born to any married couple. This is not as hypothetical as it may seem; the topic of government control has frequently been raised in other less developed countries, yet the population in Britain is more concentrated than it is in, say, India. We have been deluding ourselves for years into believing that we can manage to handle a greatly increased population, and in so doing have brainwashed the present population into accepting that a population figure of between 66 and 70 million people is unavoidable by the end of the century. This is not so, but it can only be prevented by a more realistic approach by religious bodies, the Government and individuals to the question of planned families.

The current situation in Great Britain is already quite clear to see. The birth rate exceeds the number of deaths by 300,000 per annum. If this continues unchecked, the population will increase from its current 53 million to between 66 million and 70 million by A.D. 2000. This growth can be stopped and reversed if the present average size of families is reduced from 2·5 children to 2·25. The present population exceeds what is considered to be the optimum level by something in the order of 14 million people. This is probably the last generation able to take the necessary steps to avert an intolerable situation without resorting to compulsion of some form.

The large number of illegitimate children, in addition to those conceived before marriage, and those unwanted after marriage, indicates a woeful neglect or lack of understanding of contraception methods. It should be emphasised that this is not a book on morals: it is a book dealing with the realities of the pressures on the environment and the origins of pollution. And one wonders if those who prattle on about 'no birth control' and 'no family planning' are prepared to accept the problems of overcrowding and pollution that will result from following this broad road to destruction.

It is essential that contraception is taught in all schools as part of a general course on sex education. It is in any case wrong that any child should be taught the facts of life without being informed on birth control. We are also reaching the stage where it is cheaper to provide free contraceptives than to provide for the care of unwanted children, and there may well be a clear justification for this in the near future. Certainly the voluntary sterilisation of males should become accepted as a reasonable method of population control. There is a clear case for this to be a free service provided under the National Health Service. It is cheaper than the corresponding operation for women, and over a long period much cheaper than providing free contraceptives.

From a global viewpoint population growth and food shortage are the greatest problems. It took 8,000 years for the world population to reach its first 1,000 million, a further 175 years to double that to 2,000 million, and only 55 years will be needed to double that again to 4,000 million by 1980. By the year 2000 it could be well over 6,000 million people, all wanting food, education, houses and jobs, and all contributing to problems of air pollution, waste and sewage. This is the magnitude of the problem which concerns us all. Failure to solve the problem ourselves may mean that it will be solved for us by a global war, or by the spread of disease and illness.

It is very easy for most of us now alive to ignore the increasing size of the overall problem of the environment. It is really a question of what type of life we want to leave for our children and grandchildren, or any further generations. We are now at the turning point where we are able to envisage the appalling consequences of doing nothing, and have the knowledge and ability to prevent them ever occurring. The first steps have already shown what can be achieved when the will exists. The next few years will show whether the politicians have just been paying lip service to a very real problem or if they are prepared to lead the way in passing the necessary legislation and making money available for the urgently required improvements.

The choice is ours: either a clean pleasant world with enough room to avoid treading on each other, or an endless, overcrowded, polluted conurbation. Now is the time to decide.

Appendix

The following list of addresses of some of the major organizations concerned with preserving nature and improving the quality of living are included in the hope that readers will be encouraged to find out more about those that interest them, and possibly become members.

ASSOCIATION FOR THE PRESERVATION OF RURAL SCOTLAND,
39 Castle Street, Edinburgh 2.

BRITISH ASSOCIATION FOR THE CONTROL OF AIRCRAFT NOISE,
2 Serjeants Inn, London, E.C.4.

CENTRAL COUNCIL FOR RIVERS PROTECTION,
Fishmongers' Hall, London Bridge, London, E.C.4.

CIVIC TRUST,
18 Carlton House Terrace, London, S.W.1.

CIVIC TRUST FOR WALES/TREFTADAETH CYMRU,
Department of Town Planning, University of Wales Institute of Science and Technology, Greyfriars Road, Cardiff.

SCOTTISH CIVIC TRUST,
24 George Square, Glasgow, C.2.

CIVIC TRUST FOR THE NORTH EAST,
34/35 Saddler Street, Durham.

CIVIC TRUST FOR THE NORTH WEST,
Century House, St Peter's Square, Manchester, M2 3DN.

COASTAL ANTI-POLLUTION LEAGUE,
'Alverstoke', Greenway Lane, Bath, Somerset.

CONSERVATION SOCIETY,
21 Hanyards Lane, Cuffley, Potters Bar, Hertfordshire.

COUNCIL FOR THE PROTECTION OF RURAL ENGLAND,
4 Hobart Place, London, S.W.1.

COUNCIL FOR THE PROTECTION OF RURAL WALES,
Melfod, Montgomeryshire.

INLAND WATERWAYS ASSOCIATION,
114 Regent's Park Road, London, N.W.1.

KEEP BRITAIN TIDY GROUP,
76 Strand, London, W.C.2.

MEN OF THE TREES,
Hollybank House, Emsworth, Hampshire, PO10 7UX.

NATIONAL SOCIETY FOR CLEAN AIR,
Field House, Breams Buildings, London, E.C.4.

NATIONAL TRUST,
42 Queen Anne's Gate, London, S.W.1.

NATIONAL TRUST FOR SCOTLAND,
5 Charlotte Square, Edinburgh, 2.

NOISE ABATEMENT SOCIETY,
6/8 Old Bond Street, London, W.1.

ROYAL SOCIETY FOR THE PROTECTION OF BIRDS,
The Lodge, Sandy, Bedfordshire.

UNITED KINGDOM FEDERATION AGAINST AIRCRAFT NUISANCE,
9 Dover Park Drive, London, S.W.15.

Sources and recommended reading

1 ARTICLES AND BOOKS DEALING WITH POLLUTION, THE ENVIRONMENT AND THE COUNTRYSIDE

Nature Conservation in Britain, D. Stamp (Collins, 1969)

Man and Nature (1969 Reith Lectures), F. F. Darling (B.B.C. Publications, 1970)

Conservation, J. Joffe (Aldus Books, 1969)

Journey Through Britain, J. Hillaby (Constable, 1968)

Man and Environment, R. Arvill (Penguin Books, 1967)

Pesticides and Pollution, K. Mellanby (Collins, 1967)

The Environmental Revolution, M. Nicholson (Hodder & Stoughton, 1970)

Before Nature Dies, J. Dorst (trans. C. D. Sherman) (Collins, 1970)

Environmental Conservation, R. F. Dasmann (Wiley, 2nd edn, 1968)

The Assaults On Our Senses, J. Barr (Methuen, 1970)

Cleaning Our Environment, The Chemical Basis For Action (American Chemical Society, 1969)

Restoring the Quality of Our Environment – Report of the Environmental Pollution Panel of the President's Advisory Committee (U.S. Government, 1965)

A Better Country, C. Chataway (Chairman) (Conservative Political Centre, 1966)

Royal Commission on Environmental Pollution: First Report, Sir E. Ashby (Chairman) (HMSO, 1971)

'Clean and Pleasant Land', *C.P.C.* No. 469 (Conservative Political Centre, 1970)

'Controlling Our Environment', Lord Kennet, *Fabian Research Series* No. 283 (Fabian Society, 1970)

'Pollution of the Environment', *R.I.C. Reviews* Vol. 3, No. 2 (Royal Institute of Chemistry, 1970)

God's Own Junkyard, P. Blake (Holt, Rinehart & Winston, 1964)

Dangerous Properties of Industrial Materials, N. I. Sax (ed.), 3rd edn (Reinhold, 1968)

Our Polluted Planet, Ambassador College Research Department (1968)

'The Urban Pollution Battle', *Municipal Review* (1970), Vol. 41, No. 484

Annual Report of the Scientific Adviser 1968, B. R. Brown, Greater London Council, Scientific Branch (1969)

Annual Report of the Scientific Adviser 1969, B. R. Brown, Greater London Council, Scientific Branch (1970)

'Can Man Survive?', P. Welbourn, *Weekend Magazine*, 3 January 1970

'Blighted Britain', D. Hart-Davis, 4 Parts, *Sunday Telegraph*, March–April 1970

'Social Criticism and the Chemical Industry', S. S. Castle, *Chemical and Engineering News*, p. 144, 11 October 1965

'Chemical Warfare', E. Huxley, *Punch*, p. 766, 18 November 1964

'Last Chance for Mother Earth', *Time Magazine*, 2 February 1970

2 ARTICLES AND BOOKS DEALING WITH WATER POLLUTION, WATER PURIFICATION, SEWAGE AND SEA POLLUTION

The Examination of Water and Water Supplies, E. V. Suckling, 5th edn (J. & A. Churchill, 1943)

The Examination of Water and Water Supplies, E. W. Taylor, 7th edn (J. & A. Churchill, 1958)

River Pollution, L. Klein, 3 vols (Butterworth, 1959)

Chemical Analysis of Industrial Water, J. W. McCoy (Macdonald, 1969)

Developments in Water Quality Research, H. I. Shuval (ed.) (Ann Arbor-Humphrey, 1970)

Demineralization by Electrodialysis, J. R. Wilson (ed.) (Butterworth, 1960)

Water Pollution Research 1967, Ministry of Technology (HMSO, 1968)

Water Pollution Research 1968, Ministry of Technology (HMSO, 1969)

Water Pollution Research 1969, Ministry of Technology (HMSO, 1970)

Proceedings of the 1st International Symposium on Water Desalination, 3 vols (U.S. Dept. of the Interior, 1965)

Thermal Pollution and Aquatic Life', J. R. Clark, *Scientific American*, Vol. 220, No. 3, March 1969

'When A River Becomes An Open Drain', J. Tinker, *Daily Telegraph Magazine*, No. 310, 25 September 1970

The Struggle For Clean Water, Public Health Service Publication No. 958 (U.S. Dept. of Health, Education and Welfare, 1962)

'Ground Water', A. N. Sayre, *Scientific American*, Reprint 818, November 1950

'The Ageing Great Lakes', C. F. Powers and A. Robertson, *Scientific American*, Vol. 215, No. 5, November 1966

'Spring-Clean For Britain's Rivers', J. Williams, *The Reader's Digest*, Vol. 96, p. 27, March 1970

'The Re-use of Water in the Electroplating Industry', H. Silman, *Chemistry and Industry*, p. 2046 (1962)

'Biochemical Reduction of Waste', J. R. Simpson, *Process Biochemistry*, p. 47, July (1970)

The Black Tide In The Wake of Torrey Canyon, R. Petrow (Hodder & Stoughton, 1968)

Oil and Water – The Torrey Canyon Disaster, E. Cowan (William Kimber, 1969)

The Torrey Canyon, Report of the Committee of Scientists on the Scientific and Technological Aspects of the Torrey Canyon Disaster (HMSO, 1967)

Coastal Pollution – Observations on the Report of the Select Committee on Science and Technology (HMSO, 1969)

'The Dirty Deep', M. Allaby, *The Ecologist,* p. 40, Vol. 1, No. 2 (1970)

Discharge of Sewage and Trade Effluents Into Tidal Waters, J. Inst. of Sewage Purification, p. 119, No. 2 (1964)

Sewerage and Sewage Disposal, L. B. Escritt, 3rd edn (C.R. Books, 1965)

Waste Treatment, P. C. G. Isaac (ed.) (Pergamon, 1960)

Town Waste Put To Use, P. Wix (ed.) (Cleaver-Hume, 1961)

'Sewerage In Ancient and Mediaeval Times', H. F. Gray, *Sewage Works Journal,* p. 939, Vol. 12 (1940)

'Sewage Works Design Requirements', L. G. Rice, *Water and Sewage Works,* p. 436, Vol. 97 (1950)

'The Composition of Sewage and Its Potential Use as a Source of Industrial Water', S. H. Jenkins, *Chemistry and Industry,* p. 2072, 15 December (1962)

Report On The Operation of the Services Administered by the Sewerage Board for the Year 1969, States of Jersey (1970)

3 ARTICLES AND BOOKS DEALING WITH ASPECTS OF AIR POLLUTION

Air Pollution – Its Origins and Prevention, A. R. Meetham, 3rd edn (Pergamon, 1964)

Air Pollution, M. W. Thring (ed.) (Butterworth, 1957)

Air Pollution, A. C. Stern (ed.), 3 Vols., 2nd edn (Academic Press, 1968)

Clean Air – Law and Practice, J. F. Garner and R. K. Crow, 3rd edn (Shaw and Sons, 1969)

An Introduction to Electrostatic Precipitation, H. E. Rose and A. J. Wood, 2nd edn (Constable, 1966)

Committee on Air Pollution – Interim Report, H. Beaver (Chairman) (HMSO, 1953)

Committee on Air Pollution – Final Report, H. Beaver (Chairman) (HMSO, 1954)

Measurement of Air Pollutants – Guide to the Selection of Methods, M. Katz (World Health Organisation, 1969)

Industrial Electrostatic Precipitation, H. J. White (Addison-Wesley/Pergamon, 1963)

The Effects of Air Pollution on Plants and Soil (Agricultural Research Council, 1967)

Particulate Clouds: Dusts, Smokes and Mists, H. L. Green and W. R. Lane, 2nd edn (E. & F. N. Spon, 1964)

Air Pollution Control, W. L. Faith (Wiley, 1959)

Methods of Measuring Air Pollution (Organisation for Economic Co-operation and Development, 1964)

Fumifugium or The Inconvenience of the Aer and Smoake of London Dissipated together with some Remedies Humbly proposed, J. Evelyn, published 1661 (reprinted by the Oxford University Press, 1930)

'On the Influences of Changes in the CO_2 Concentration in Air on the Radiation Balance of the Earth's Surface and on the Climate', F. Möller, *J. Geophysical Res.*, p. 3877, Vol. 68 (1963)

104th Report on Alkali, & C. Works, 1967, Chief Inspectors (HMSO, 1968)

105th Report on Alkali, & C. Works, 1968, Chief Inspectors (HMSO, 1969)

106th Report on Alkali, & C. Works, 1969, Chief Inspectors (HMSO, 1970)

'Content to Breath', R. Allen, *The Ecologist*, p. 38, Vol. 1, No. 2 (1970)

'The Global Village Pump', F. Wheeler, *New Scientist,* p. 10, No. 721, Vol. 48 (1970)

'The Control of Air Pollution', A. J. Haagen-Smit, *Scientific American,* No. 1, Vol. 210, January (1964)

'Air Pollution and Public Health', W. McDermott, *Scientific American,* Reprint 612, October (1961)

'Wage War On Smog', *C.P.C.* No. 202 (Conservative Political Centre, 1959)

'Carbon Dioxide and Climate', G. N. Plass, *Scientific American,* Reprint 823, July (1959)

4 ARTICLES AND BOOKS DEALING WITH THE CREATION AND DISPOSAL OF SOLID WASTES

The Waste Makers, V. Packard (Penguin Books, 1963)

Public Cleansing, F. Flintoff and R. Millard (MacLaren & Sons, 1969)

Disposal of Solid Toxic Wastes, A. Key (Chairman), Ministry of Housing and Local Government (HMSO, 1970)

Disposal of Plastics Waste and Litter, J. J. P. Staudinger, Monograph No. 35 (Society of Chemical Industry, 1970)

Atmospheric Oxidants and Antioxidants, G. Scott (Elsevier, 1965)

'Vanishing Plastics', G. Scott, *New Scientist,* p. 293, Vol. 47, No. 713 (1970)

'Waste Disposers', Consumers' Association, *Which?,* p. 46 (1968)

Third Pollution – The National Problem of Solid Waste Disposal, W. E. Small (1970)

'Plastics In Packaging', C. T. Coward, *Chemistry In Britain,* p. 276, Vol. 6, No. 6 (1970)

5 ARTICLES AND BOOKS ON THE USE AND MISUSE OF LAND

Derelict Britain, J. Barr (Penguin Books, 1969)

Silent Spring, R. Carson (Houghton Mifflin, 1962)

Since Silent Spring, F. Graham (Houghton Mifflin, 1970)

Degradation of Herbicides, P. C. Kearney and D. D. Kaufman (eds) (Marcel Dekker, 1969)

Aberfan: The Story of a Disaster, T. Austin (Hutchinson, 1967)

Forestry in the British Scene, Forestry Commission Booklet No. 24 (HMSO, 1968)

Sylva, – or a discourse of Forest-trees and the Propagation of Timber in His Majesties Dominions, J. Evelyn (Royal Society of London, 1664)

Pest Control Without Poisons, L. D. Hills (Henry Doubleday Research Association, 1964)

'Forestry in Kent', Forestry Commission, *Defend Kent,* p. 10, No. 3 (1970)

'Pesticides and the Reproduction of Birds', D. B. Peakall, *Scientific American,* p. 73 (1970)

Pesticide Residues, J. Thomson and D. C. Abbott (Royal Institute of Chemistry, 1966)

Chemistry and Crop Protection, R. L. Wain (Royal Institute of Chemistry, 1965)

Agriculture and Some of Its Chemical Problems, E. G. Cox (Royal Institute of Chemistry, 1963)

'Landscaping Motorways', A. P. Dunball, *Defend Kent,* p. 6, No. 3 (1970)

'Newtown, Exeter', M. Hook, *Architects' Journal,* p. 1488 (1970)

'Burley Lodge, Leeds', M. Hook, *Architects' Journal,* p. 1464 (1970)

'Putting Power Stations Where They Belong', *Nature,* p. 987, Vol. 227 (1970)

'Soil Pollutants and Soil Animals', C. A. Edwards, *Scientific American,* p. 88, Vol. 220, No. 4 (1969)

'Toxic Substances and Ecological Cycles', G. M. Woodwell, *Scientific American,* p. 24, Vol. 216, No. 3 (1967)

'Third-Generation Pesticides', C. M. Williams, *Scientific American,* p. 13, Vol. 217, No. 1 (1967)

6 PUBLICATIONS ON RADIOACTIVITY AND RADIATION PROTECTION

Radioactive Wastes: Their Treatment and Disposal, J. C. Collins (ed.) (E. & F. N. Spon, 1960)

Atomic Radiation and Life, P. Alexander (Penguin Books, 1957)

Perils of the Peaceful Atom, R. Curtis and E. Hogan (Victor Gollancz, 1970)

Law Relating to Nuclear Energy, H. Street and F. R. Frame (Butterworth, 1966)

Principles of Radiation Protection, G. Eaves (Iliffe, 1964)

Electricity Without Dynamos, J. W. Gardner (Penguin Books, 1963)

The Voyage of the Lucky Dragon, R. E. Lapp (Penguin Books, 1957)

Code of Practice For The Protection of Persons Exposed to Ionising Radiations in Research and Teaching, Department of Employment and Productivity (HMSO, 1968)

Reactor Safeguards, C. R. Russell (Pergamon, 1962)

Nuclear Energy in Britain, Central Office of Information, (HMSO, 1957)

Living With The Atom, R. Calder (University of Chicago Press, 1962)

The Atom and the Energy Revolution, N. Lansdell (Penguin Books, 1958)

Principles of Radiation Protection, K. Z. Morgan and J. E. Turner (eds) (Wiley, 1967)

'Ionizing Radiation', Special Issue, *Scientific American*, Vol. 201, No. 3 (1959)

'Radiation and Human Mutation', H. J. Muller, *Scientific American*, Reprint No. 29, November 1955

'Is There A Peaceful Atom?', P. Bunyan, *The Ecologist*, p. 15, Vol. 1, No. 1 (1970)

'Regulations for the Safe Transport of Radioactive Materials',

Safety Series No. 6, International Atomic Energy Agency (1967)

The Natural Radiation Environment, J. A. S. Adams and W. M. Lowderd (eds) (University of Chicago Press, 1964)

Radiation Control For Fire and Other Emergency Forces, A. A. Keil (National Fire Protection Association (U.S.A.), 1960)

First International Symposium on the Decontamination of Nuclear Installations, H. J. Blythe (ed.) (Cambridge University Press, 1967)

Radioactive Waste Control at the United Kingdom Atomic Research Establishment, Harwell, R. H. Burns, G. W. Clare and J. H. Clarke (HMSO, 1968)

Accident at Windscale No. 1 Pile on 10th October, 1957, W. Penney (Chairman) (HMSO, 1957)

'Radiation and the Human Cell', T. T. Puck, *Scientific American*, p. 142, Vol. 202, No. 4 (1960)

Final Report of the Committee Evaluating the Design and Operation of the Windscale Piles, A. Fleck (Chairman) (HMSO, 1958)

7 BOOKS AND PUBLICATIONS ON SOUND AND NOISE CONTROL

Noise and Society, M. Rodda (Oliver and Boyd, 1967)

Noise Measurement and Control, P. Lord and F. L. Thomas (Heywood and Co., 1963)

Noise Reduction, L. L. Beranek (ed.) (McGraw-Hill, 1960)

The Measurement and Suppression of Noise, A. J. King (Chapman and Hall, 1965)

Technical Aspects of Sound, E. G. Richardson and E. Meyer (eds), 3 vols (Elsevier, 1962)

Noise – Final Report, A. Wilson (Chairman) (HMSO, 1963)

Handbook of Noise Control, C. M. Harris (ed.) (McGraw-Hill, 1957)

Acoustic Noise Measurements, J. T. Broch (Brüel and Kjaer, 1970)

Action Against Aircraft Noise, Board of Trade (HMSO, 1969)

'Defeating The Decibels', B. Goodson, *Do It Yourself Annual*, p. 41 (Link House Publications, 1969)

'Beating Noise', B. Goodson, *Do It Yourself Magazine* (Link House Publications, September and October, 1970)

'Effects of Jet Aircraft Noise On Hearing', R. R. A. Coles and J. J. Knight, *Nature*, p. 1803, Vol. 184 (1959)

'The Menace of the Boom', B. Lundberg, *The Observer*, 5 June (1966)

'The Ear', G. von Bekesy, *Scientific American*, p. 66, Vol. 197, No. 2 (1957)

'Noise Control', P. H. Smith, *Scientific American*, p. 204, Vol. 161 (1939)

'Recent Developments in Vocal Acoustics', E. G. Richardson, *Discovery*, p. 333, Vol. 21 (1958)

8 PUBLICATIONS CONCERNED WITH POPULATION GROWTH AND PRESSURES ON THE ENVIRONMENT

Population Growth and Land Use, C. Clark (Macmillan, 1967)

New Lives, New Landscapes, N. Fairbrother (The Architectural Press, 1970)

'World Population', J. Huxley, *Scientific American*, p. 64, Vol. 194, No. 3 (1956)

'The Human Population', E. S. Deevey, *Scientific American*, p. 195, Vol. 203, No. 3 (1960)

The Optimum Population For Britain, L. R. Taylor (ed.) (Academic Press, 1970)

'Population', K. Davis, *Scientific American*, p. 63, Vol. 209, No. 3 (1963)

The Crowded World, J. Huxley, 'Your Environment', p. 125, No. 4 (1970)

Index